Notes on logic and set theory

P. T. JOHNSTONE

University Lecturer in Pure Mathematics
University of Cambridge

CAMBRIDGE
UNIVERSITY PRESS

Published by the Press Syndicate of the University of Cambridge
The Pitt Building, Trumpington Street, Cambridge CB2 1RP
40 West 20th Street, New York, NY 10011–4211, USA
10 Stamford Road, Oakleigh, Melbourne 3166, Australia

First published 1987
Reprinted 1992, 1996

British Library cataloguing in publication data
Johnstone, P.T.
Notes on logic and set theory.
1. Logic, symbolic and mathematical
I. Title
511.3   BC135

Library of Congress cataloguing in publication data
Johnstone, P. T.
Notes on logic and set theory.
1. Logic, Symbolic and mathematical.   2. Set theory.
I. Title.
QA9.J64   1987   511.3   87-11758

ISBN 0 521 33502 7 hard covers
ISBN 0 521 33692 9 paperback

Transferred to digital printing 2002

Notes on logic and set theory

Contents

Preface

This book has its origins in a course of lectures entitled 'Set Theory and Logic' which is given to third-year undergraduates in Cambridge. The Cambridge Mathematical Tripos contains rather little on the foundational aspects of mathematics: this course is (at the time of writing – though there are plans for a change before long) the only opportunity which undergraduates have to learn about the basic ideas of logic and axiomatic set theory in an examinable context, and its aim is therefore to provide a general introduction to these ideas for students who, though they may possess considerable sophistication in other areas of mathematics, have no previous experience of logic beyond a nodding acquaintance with 'naive' set theory, and whose primary interests may well lie in other areas of pure (or even applied) mathematics. Having lectured this course in 1984, 1985 and 1986, I have been struck by the fact that there was no single textbook available covering both logic and set theory at the level of rigour and sophistication appropriate to such a course, and – in the belief that there might be universities other than Cambridge where the need for such a textbook was felt – I conceived the idea of expanding my lecture notes into a publishable text. I am glad to say that this idea was enthusiastically received by Cambridge University Press; I am grateful, in particular, to David Tranah and Martin Gilchrist for their support of the project.

The *raison d'être* of this book, then, is to try to collect within a single pair of covers everything that the well-educated mathematician in the late twentieth century needs to know about the foundations of his subject. Though there has, as indicated above, been some expansion as compared with the original lecture course, it has been kept to a bare minimum: anything which is merely of specialist interest to logicians has (I hope) been excluded. (Briefly, the additions consist of the whole of Chapter 4 on recursion theory – there was, regrettably, not enough time to cover recursive functions in the original course – plus more detailed proofs of one or two major results (such as the Completeness Theorem for the Predicate Calculus) which were merely stated, or given sketch-proofs, in the lectures.)

However, I have refrained from giving the book a title of the *Logic for the Working Mathematician* variety, since such titles are often a cover for compromise: in their anxiety to make material accessible to the general reader, authors have a tendency to skate over details when to do otherwise would involve them in discussion of what might be considered specialist matters. (I should hastily add that the last sentence is not intended as a criticism of any particular book whose title may approximate to that quoted.) In this book I have tried to be completely honest with the reader; I have always sought the simplest and most direct proof, but where there are genuine technical difficulties in the way I have not been afraid to say so (even though, in some cases, it has not been possible within the compass of a book like this to give a full account of the difficulties or of the manner of their resolution). In an effort to prevent the text from becoming too indigestible, I have often relegated the discussion of these difficulties to a remark (sometimes headed 'Note for worriers', or something equally facetious) enclosed within square brackets; the reader who wishes to confine himself to the basic ideas and not bother about the technicalities can usually skip over such bracketed passages without loss of continuity.

The layout of the book is as follows. Chapters 1–3 develop the language and machinery of first-order logic, up to the proof of the Completeness Theorem and some of its consequences. Chapter 4 develops recursion theory from its beginnings up to (but not quite including) the Recursion Theorem. Chapters 5–8 develop Zermelo–Fraenkel set theory, beginning with the axioms and working up to the 'traditional' discussion of ordinal and cardinal arithmetic. Finally, Chapter 9 contains a proof of Gödel's Incompleteness Theorems, followed by a fairly informal discussion of the technology of set-theoretic independence proofs. There are exercises at the end of each chapter (except Chapter 9, where it did not seem appropriate);

these range from mere five-finger exercises, through the verification of minor details which have been omitted from the proofs of theorems in the text, to quite substantial problems whose results might well have been included in the text as propositions, had it not been for the policy of exclusion mentioned earlier. The latter kind (in particular) have been provided with what I hope is a generous supply of hints, enclosed within square brackets.

Although it is clearly preferable that a student should work through the whole book from beginning to end, it is possible to meet the demands of shorter courses by omitting material in various ways. In particular, if it is desired to omit recursion theory (as in the course from which the book developed), then Chapter 4 can be skipped without any loss of understanding of Chapters 5–8; there will inevitably be more difficulty with the Incompleteness Theorems in Chapter 9, but I hope that it would still be possible to understand the discussion of them on an informal level. On the other hand, if one is not particularly interested in covering axiomatic set theory, it should be possible to jump from the end of Chapter 4 – or, even better, from about halfway through Chapter 5 – to the beginning of Chapter 9.

The prerequisites for reading the book are fairly few; in particular, though the opening chapters presuppose some acquaintance with naive set theory, there is no presumption that the reader knows anything about logic. Otherwise, his mathematical knowledge is assumed to be what would be usual for a final-year undergraduate in Britain: some familiarity with subjects such as group theory, ring theory and point-set topology is presumed in some of the examples, but the omission of these examples should not seriously damage one's understanding of the mainstream of the text. Thus, although this is not its primary purpose, the book could be used as a first course in logic for rather more specialist logicians than I have envisaged.

There is one respect, however, in which the book is rather conspicuously aimed at non-specialists: namely, I have not been afraid to shoot a few of the sacred cows of logical tradition for the sake of rendering the exposition more smoothly compatible with mathematical practice. Perhaps the most obvious instance occurs in Chapter 3 where, in order to develop a logical calculus which is both sound and complete for possibly-empty models, I found it necessary to introduce the restriction that quantifiers may only be applied to formulae which actually involve the quantified variable [if $(\forall x)\bot$

were allowed as a formula, then neither of the implications $((\forall x)\bot \Rightarrow (\forall x)\neg(x = x))$ and $((\forall x)\neg(x = x) \Rightarrow (\forall x)\bot)$ would be provable]. I can well imagine that professional logicians will find this hard to swallow (I did myself at first, though I am not by training a logician); but I would ask them to reflect, before they condemn me, on whether anyone *other* than a logician would find this restriction at all odd.

Although, as indicated in the first paragraph, the selection of topics covered in this book does not exactly correspond to that in any previous book known to me, there are a number of previous texts from which I have borrowed ideas concerning the arrangement and presentation of material, and to whose authors I must therefore express my indebtedness – it would be obvious in any case from the text itself. For the logic chapters (1–3), I have taken various ideas from *An Algebraic Introduction to Mathematical Logic* by D. W. Barnes and J. M. Mack (Springer-Verlag, 1975); in particular, the idea of beginning a treatment of first-order logic by studying universal algebra is theirs. In Chapter 4, the influence of N. J. Cutland's book *Computability* (Cambridge University Press, 1980) will be apparent to anyone who knows it well. And in the set theory chapters (5–8), the ordering of material (in particular) owes a good deal to *Sets: Naive, Axiomatic and Applied* by D. van Dalen, H. C. Doets and H. de Swart (Pergamon Press, 1978). [The citation of these three texts has an ulterior motive. It occurs to me that many students, having read the deliberately brief accounts of first-order logic, recursion theory and set theory in this book, will want to pursue at least one of them in greater depth. For this purpose, I should be happy to steer them towards the three books just mentioned.]

But, far more than to any of these authors, I am indebted to the three generations of undergraduates who struggled through the lectures that generated successive approximations to these notes, and to my colleagues who, in tutorials, found themselves obliged to patch up my mistakes and eliminate the misunderstandings I had created. It would be invidious to mention any particular names, but it is undoubtedly true that, but for the feedback I received from them, these notes would contain a great many more errors than they do.

Cambridge, January 1987   P. T. Johnstone

1

Universal algebra

The function of mathematical logic is to provide formal languages for describing the structures with which mathematicians work, and the methods of proof available to them. Obviously, the more complicated we make our language, the more powerful it will be as an instrument for expressing our ideas; but in these notes we are going to begin with what is perhaps the simplest *useful* language, that of universal algebra (sometimes called equational logic). Although simple, it displays many of the characteristic features of more complicated languages, which is why it makes a good introduction to them.

Universal algebra begins by abstracting the common features of a number of familiar mathematical structures, including groups, rings and vector spaces. In each of these cases, the structure is defined to be a set equipped with certain finitary *operations*, which satisfy certain *equations*. For example, a group is a set G equipped with

a binary operation $m: G \times G \to G$ (multiplication),
a unary operation $i: G \to G$ (inversion),
and a nullary operation $e: G^0 \to G$ (identity)

[note: we adopt the convention that G^0 is a singleton set for any G, so that a 0-ary operation – also called a constant – simply picks out a single element of G], satisfying the equations

$m(x, m(y, z)) = m(m(x, y), z)$ (associative law),
$m(e, x) = x$ (left identity law),
and $m(i(x), x) = e$ (left inverse law),

which are supposed to hold for all possible values of x, y, z in G.

We leave it as an exercise for the reader to write down similar descriptions of the notion of ring (with 1), and of the notion of K-vector space for a given field K. Note that for the latter it is necessary (or at least convenient) to take (scalar multiplication by) each element of K as a unary operation; thus in general the set of operations and/or equations required to define a given type of structure may be infinite.

Abstracting from the above examples, we introduce the notion of an operational type. An *operational type* is a pair (Ω, α) where Ω is a set of *operation-symbols* and α is a function assigning to each $\omega \in \Omega$ a natural number $\alpha(\omega)$, called its *arity*. [N.B.: throughout these notes, 0 is considered to be a natural number.] Frequently, we suppress any explicit mention of the function α, and simply write 'Ω is an operational type'. Thus in our example above we have $\Omega = \{m, i, e\}$ with $\alpha(m) = 2$, $\alpha(i) = 1$, $\alpha(e) = 0$.

Given an operational type (Ω, α), a *structure* of type (Ω, α) (or Ω-structure, or Ω-algebra) is a set A equipped with functions $\omega_A : A^{\alpha(\omega)} \to A$ for each $\omega \in \Omega$. We call ω_A the *interpretation* of the abstract symbol ω in the structure A; we also speak of the family of functions $(\omega_A \mid \omega \in \Omega)$ as an Ω-structure on the set A. A *homomorphism* $f : A \to B$ of Ω-structures is a function such that

$$f(\omega_A(a_1, \ldots, a_{\alpha(\omega)})) = \omega_B(f(a_1), \ldots, f(a_{\alpha(\omega)}))$$

for all $\omega \in \Omega$ and all $a_1, a_2, \ldots, a_{\alpha(\omega)}$ in A.

So much for the operations; how about the equations? Before answering this question, we turn to a seemingly different topic: the notion of a *term* or *derived operation*. Let Ω be an operational type and X a set (whose elements we shall call *variables*; we assume for convenience $\Omega \cap X = \varnothing$); then the set $F_\Omega(X)$ (or simply FX) of Ω-*terms* in X is defined inductively as follows:

(a) If $x \in X$, then $x \in F_\Omega(X)$.
(b) If $\omega \in \Omega$, $\alpha(\omega) = n$ and $t_1, t_2, \ldots, t_n \in F_\Omega(X)$, then $\omega t_1 t_2 \ldots t_n \in F_\Omega(X)$.
(c) That's all.

[Inductive definitions of this type are very common in the construction of formal languages. Formally, what this one means is that $F_\Omega(X)$ is the smallest subset of the set $(M$, say) of all finite strings of elements of $\Omega \cup X$ which satisfies the closure properties (a) and (b), i.e. the intersection of all such subsets.]

Remark 1.1. Note in passing that we have a simple algorithm for determining whether a given finite string in M belongs to $F_\Omega(X)$: start at the right-hand end of the string with counter set to 0, and move leftwards increasing the count by 1 each time you pass a variable, and decreasing it by $n-1$ each time you pass an n-ary operation-symbol. Then the string belongs to FX iff the counter never falls below 1 after your start, and finishes at 1. As illustrations, we show the counter values for a number of strings where Ω is the operational type of groups and $X = \{x, y, z\}$; the first one is in FX, the other two are not.

$$1\ 2\ 1\ 2\ 3\ 3\ 2\ 1\ 1\ 0 \qquad 1\ 2\ 2\ 2\ 1\ 0\ 1\ 0 \qquad 3\ 3\ 2\ 3\ 2\ 2\ 1\ 0$$
$$m\ e\ m\ m\ i\ x\ y\ i\ z\ ; \qquad m\ i\ i\ x\ y\ m\ z\ ; \qquad i\ x\ m\ e\ i\ y\ x\ .$$

Theorem 1.2. (i) $F_\Omega(X)$ has an Ω-structure.

(ii) $F_\Omega(X)$ is the *free* Ω-structure generated by X; i.e., given any Ω-structure A and any function $f: X \to A$, there exists a unique homomorphism $\bar{f}: F_\Omega(X) \to A$ extending f.

Proof. The existence of the Ω-structure is immediate from clause (b) of the definition: if $\omega \in \Omega$ (with $\alpha(\omega) = n$, say) and $t_1, t_2, \ldots, t_n \in FX$, define

$$\omega_{FX}(t_1, t_2, \ldots, t_n) = \omega t_1 t_2 \ldots t_n$$

(i.e. just erase the brackets and commas).

Part (ii) is essentially a matter of putting the brackets back in. Since FX was defined inductively, we can define $\bar{f}$ inductively:

if $t = x \in X$, then $\bar{f}(t) = f(x)$;
if $t = \omega t_1 \ldots t_n$ where $\omega \in \Omega$, $\alpha(\omega) = n$ and $\bar{f}$ has already been defined at $t_1, \ldots, t_n \in FX$, then $\bar{f}(t) = \omega_A(\bar{f}(t_1), \ldots, \bar{f}(t_n))$.

It is then clear that $\bar{f}$ is a homomorphism, and that it is the unique homomorphism extending f. $\square$

Another important (and trivial) property of free Ω-structures is

Lemma 1.3. For any X,

$$F_\Omega(X) = \bigcup \{F_\Omega(X') | X' \subseteq X, \ X' \text{ finite}\}. \quad \square$$

Thus we may, for many purposes, restrict our attention to free structures generated by finite sets. Let $X_n = \{x_1, x_2, \ldots, x_n\}$ be a standard n-element set; let $t \in FX_n$, and let A be any Ω-structure.

Then we may define a function $t_A: A^n \to A$ inductively as follows:

if $t = x_i$ $(1 \leqslant i \leqslant n)$, then t_A is projection onto the ith factor;

if $t = \omega t_1 t_2 \ldots t_m$ where $\alpha(\omega) = m$, then t_A is the composite

$$A^n \xrightarrow{((t_1)_A, \ldots, (t_m)_A)} A^m \xrightarrow{\omega_A} A.$$

In particular, if t is the term $\omega x_1 x_2 \ldots x_n$, where $\alpha(\omega) = n$, then $t_A = \omega_A$. The function t_A is called (the interpretation in A of) the n-ary *derived operation* corresponding to the term t (in contrast to the 'primitive operations' which are the functions of the form ω_A). It is easy to see that a homomorphism $f: A \to B$ of Ω-structures commutes with all derived operations as well as with primitive ones.

Now let us return to the equations. If we look, for example, at the associative law for groups, we see that each side of the equation is a ternary derived operation (let us call the corresponding terms s and t); and the assertion that the associative law holds in a group G is just the assertion that the functions s_G and t_G are equal. We thus define an n-ary *equation* (in an operational type Ω) to be an expression $(s = t)$, where s and t are elements of $F_\Omega(X_n)$, and we say an equation $(s = t)$ is *satisfied* in a structure A if $s_A = t_A$. Finally, we define an *algebraic theory* to be a pair $T = (\Omega, E)$ where Ω is an operational type and E is a set of equations in Ω, and we define a *model* for T (or T-algebra) to be an Ω-structure which satisfies all the equations in E.

Thus, for example, a group is exactly an (Ω, E)-model, where $\Omega = \{m, i, e\}$ as before and

$$E = \{(mx_1 mx_2 x_3 = mmx_1 x_2 x_3), (mex_1 = x_1), (mix_1 x_1 = e)\}.$$

[Note that, as in the third member of E above, it is not necessary for each of the variables $x_1, \ldots, x_n$ to appear explicitly on each side of an n-ary equation.]

Just as we did with operations, we may now enlarge the set E of 'primitive' equations to a larger set $\tilde{E}$ of derived equations. [For example, one proves in a first course on group theory that any Ω-structure satisfying the three equations in the particular E above also satisfies the 'right identity' and 'right inverse' equations $(mx_1 e = x_1)$, $(mx_1 ix_1 = e)$.] Once again, we give an inductive definition of $\tilde{E}$:

(a) $E \subseteq \tilde{E}$.

(b) $\tilde{E}$ is an equivalence relation on the set of terms: thus

    (i) for any term t, $(t = t) \in \tilde{E}$;

(ii) if $(s = t) \in \tilde{E}$, then $(t = s) \in \tilde{E}$;

(iii) if $(s = t)$ and $(t = u)$ are in $\tilde{E}$, then $(s = u) \in \tilde{E}$.

(c) $\tilde{E}$ is closed under substitution, in two different ways:

(i) if $(s = t) \in \tilde{E}$, x_i is a variable involved in s and/or t and u is any term, then $(s[u/x_i] = t[u/x_i]) \in \tilde{E}$, where $s[u/x_i]$ denotes the effect of replacing each occurrence of x_i in s by the term u;

(ii) if s is a term, x_i a variable involved in s and $(t = u)$ is in $\tilde{E}$, then $(s[t/x_i] = s[u/x_i]) \in \tilde{E}$.

(d) That's all.

[As before, this definition really means that $\tilde{E}$ is the smallest subset of the set of all expressions $(s = t)$ which is closed under (a), (b) and (c).]

If s and t are elements of $F_\Omega(X)$ for some X, let us write $s \sim_E t$ to mean $(s = t) \in \tilde{E}$; then by (b) above $\sim_E$ is an equivalence relation, and we can form the set $F_{(\Omega, E)}(X)$ of $\sim_E$-equivalence classes.

Theorem 1.4. (i) $F_{(\Omega, E)}(X)$ inherits an Ω-structure from $F_\Omega(X)$, and it satisfies the equations in E.

(ii) $F_{(\Omega, E)}(X)$ is the free (Ω, E)-model generated by X.

Proof. (i) Clause (c)(ii) of the definition of $\tilde{E}$ says that the interpretations in $F_\Omega(X)$ of the operations of Ω respect the equivalence relation $\sim_E$, and hence induce operations on the quotient set $F_{(\Omega, E)}(X)$. The fact that these induced operations satisfy the equations in E follows from ((a) and) (c)(i), since every element of $F_{(\Omega, E)}(X)$ is the equivalence class of some term.

(ii) Let $\hat{E}$ denote the set of expressions $(s = t)$ where s and t are elements of $F_\Omega(X)$ such that $h(s) = h(t)$ for every Ω-homomorphism h from $F_\Omega(X)$ to an (Ω, E)-model A. Then it is easily verified that $\hat{E}$ satisfies the closure properties (a), (b) and (c) [for (c), this requires the observation that $h(s[u/x_i]) = h'(s)$, where h' is the unique homomorphism sending x_i to $h(u)$ and the other elements of X to their images under h]; so $\tilde{E} \subseteq \hat{E}$, and hence every homomorphism $h: F(X) \to A$ factors through the quotient map $F_\Omega(X) \to F_{(\Omega, E)}(X)$. In particular, if $h = \bar{f}$ is the unique homomorphism extending a given map $f: X \to A$ (as in Theorem 1.2(ii)), we obtain a homomorphism $\bar{f}: F_{(\Omega, E)}(X) \to A$, which is clearly the unique homomorphism extending f. □

Corollary 1.5. Let (Ω, E) be an algebraic theory. Then an equation $(s = t)$ belongs to $\tilde{E}$ iff it is satisfied in every (Ω, E)-model.

Proof. One direction is easy: the set of equations satisfied in a given (Ω, E)-model (and hence, the set of equations satisfied in every (Ω, E)-model) has the closure properties (a), (b) and (c), and so contains $\tilde{E}$. Conversely, if $(s = t)$ is satisfied in every (Ω, E)-model, then it is satisfied in $F_{(\Omega, E)}(X_n)$ for any n; in particular (assuming for notational convenience that both s and t involve exactly the variables $x_1, x_2, \ldots, x_n$), we have

$$s_{F_{(\Omega, E)}(X_n)}([x_1], \ldots, [x_n]) = t_{F_{(\Omega, E)}(X_n)}([x_1], \ldots, [x_n]) \qquad (*)$$

(where the square brackets denote $\sim_E$-equivalence classes). But by definition we have

$$s_{F_{(\Omega, E)}(X_n)}([x_1], \ldots, [x_n]) = [s_{F_\Omega(X_n)}(x_1, \ldots, x_n)] = [s],$$

and similarly the right-hand side of $(*)$ equals $[t]$; so $[s] = [t]$, i.e. $(s = t) \in \tilde{E}$. $\square$

Corollary 1.5 is our first example of a *completeness theorem*, i.e. a theorem asserting (for some class of theories) that the things which are *true* (i.e. are satisfied in every model of a given theory) coincide with the things which are *provable* (i.e. are derivable from the postulates of the theory – in this case, the primitive equations – by a specified deduction process – in this case, the closure properties (b) and (c)). Clearly, the acid test of any formal deduction-system is whether we can prove a completeness theorem for it. The existence of free models, as we have seen, makes the completeness theorem for algebraic theories comparatively easy to prove; in the next two chapters we shall prove completeness theorems in other contexts where we have to do a good deal more work to show that every true statement is provable.

However, even for algebraic theories not everything in the garden is rosy. In contrast to the situation for terms, there is in general no algorithm for determining whether or not a given equation $(s = t)$ is derivable from a given theory. For some particular theories (e.g. that of groups – see Exercise 1.6) we can find such an algorithm; but in Chapter 4 we shall give an explicit example of an algebraic theory for which we can prove that no such algorithm exists. The problem of finding such an algorithm, for a given T, is called the *word problem* for T. ['Word' is an old-fashioned synonym for 'term'.]

There is one case where the word problem always has a trivial solution. Let Ω be an operational type, and let $\mathscr{A} = \{A_1, A_2, \ldots, A_n\}$ be a finite set of finite Ω-structures. Then if we define E to be the set of all equations which are satisfied in every A_i, it is clear that we already have $E = \tilde{E}$; and so to determine whether $(s = t)$ is a (derived) equation of this theory it suffices to compute s_{A_i} and t_{A_i} for each i – which is a finite process since each A_i is finite.

An important example of a theory of this kind is the theory of *Boolean algebras*, which may be loosely described as 'everything you can say about a two-element set' (that is, if you confine yourself to the language of universal algebra). There are various ways of presenting this theory: a highly generous one uses two nullary operations $\top$ (true) and $\bot$ (false), a unary operation $\neg$ (not), and four binary operations $\wedge$ (and), $\vee$ (or), $\Rightarrow$ (implies) and $\Leftrightarrow$ (iff). The set $2 = \{0, 1\}$ is given a structure for this operational type by setting

$$\top_2 = 1$$
$$\bot_2 = 0$$
$$\neg_2(a) = 1 - a$$
$$\wedge_2(a, b) = \min\{a, b\}$$
$$\vee_2(a, b) = \max\{a, b\}$$
$$\Rightarrow_2(a, b) = 0 \text{ iff } a = 1 \text{ and } b = 0$$
$$\Leftrightarrow_2(a, b) = 1 \text{ iff } a = b.$$

We then define a *Boolean algebra* to be an (Ω, E)-model, where Ω is as above and E is the set of all equations satisfied in 2. [Note: henceforth we shall generally revert to the more familiar 'algebraic' way of writing binary operations: $(x \wedge y)$ instead of $\wedge xy$, etc.] Of course, the above presentation is highly inefficient, because E contains a good many equations which tell us that some of the seven primitive operations are definable in terms of the others. For example, it is easy to verify that E contains

$$(\top = (\bot \Rightarrow \bot))$$
$$(\neg x = (x \Rightarrow \bot))$$
$$((x \vee y) = (\neg x \Rightarrow y))$$
$$((x \wedge y) = \neg(\neg x \vee \neg y))$$

and $\qquad ((x \Leftrightarrow y) = ((x \Rightarrow y) \wedge (y \Rightarrow x))),$

so that every Ω-term is $\sim_E$-equivalent to one involving only the primitive operations $\perp$ and $\Rightarrow$. Henceforth, we shall regard $\perp$ and $\Rightarrow$ as the only primitive operations in the theory of Boolean algebras, and regard the above equations as *defining* $\top$, $\neg$, $\vee$, $\wedge$ and $\Leftrightarrow$ as (shorthand for) certain derived operations. There are many other ways of reducing the number of primitive operations; this one has the (small) merit that it gets the number down to the least possible (see Exercise 1.10).

This reduction has not exhausted all the equations in E; there are still others that we need to consider. We note, however, that $(s = t)$ belongs to E iff $((s \Leftrightarrow t) = \top)$ does; therefore we can restrict our attention to equations of the form $(t = \top)$. We say a term t is a *tautology* if $(t = \top)$ is in E (equivalently, if t_2 is the constant function $2^n \to 2$ with value 1, where n is the number of variables in t). It is easy to verify that the following are tautologies:

(a) $(x \Rightarrow (y \Rightarrow x))$,
(b) $((x \Rightarrow (y \Rightarrow z)) \Rightarrow ((x \Rightarrow y) \Rightarrow (x \Rightarrow z)))$,
(c) $(((x \Rightarrow \perp) \Rightarrow \perp) \Rightarrow x)$.

(c) looks more familiar if we write it as $(\neg\neg x \Rightarrow x)$; but we wanted to emphasize that $\Rightarrow$ and $\perp$ are now our only primitive operations. We shall be meeting these three tautologies quite frequently in future.

Exercises

1.1. Let $\Omega = \{t, b, u, c\}$ with $\alpha(t) = 3$, $\alpha(b) = 2$, $\alpha(u) = 1$, $\alpha(c) = 0$, and let x, y, z be variables. Which of the following are Ω-terms?
(i) *ttxbucyzzz* (ii) *xubytcz* (iii) *tcucbucc*
(iv) *bbbxbybyybxbzbyyy* (v) *bxytczuz* (vi) *tbxxxxx*.

1.2. Show that the following definition of the derived operation induced by a term is equivalent to the one given in the text:
'If $t \in F_\Omega(X_n)$ and $a_1, \ldots, a_n$ are elements of an Ω-structure A, then $t_A(a_1, \ldots, a_n) = \bar{f}(t)$, where $\bar{f}: F_\Omega(X_n) \to A$ is the unique homomorphism extending the map $f: X_n \to A$ with $f(x_i) = a_i$ $(1 \leqslant i \leqslant n)$.'

1.3. Let s, t and u be Ω-terms (for some fixed Ω), and let x_i and x_j be distinct variables. We write $s[t, u/x_i, x_j]$ for the term obtained from s on simultaneously replacing each occurrence of x_i by t and each occurrence of x_j by u. Show that $s[t, u/x_i, x_j]$ is not in general the same as $s[t/x_i][u/x_j]$, but that it is the same as $s[t[x_n/x_j]/x_i][u/x_j][x_j/x_n]$, provided n is chosen so large that x_n does not occur anywhere in s, t or u. Hence show that if $(s = s')$, $(t = t')$ and $(u = u')$ are all derived equations of some theory (Ω, E), so is $(s[t, u/x_i, x_j] = s'[t', u'/x_i, x_j])$.

1.4.   Let T be an algebraic theory. Show that the one-element set $\{0\}$ has a unique T-model structure, and that the empty set has a T-model structure iff T contains no nullary operations.

1.5.   Let $\Omega = \{m, i, \bar{e}\}$ with $\alpha(m) = 2$, $\alpha(i) = \alpha(\bar{e}) = 1$, and let E consist of the four equations $(mxmyz = mmxyz)$, $(\bar{e}x = \bar{e}y)$, $(m\bar{e}xx = x)$ and $(mixx = \bar{e}x)$. Show that every group is an (Ω, E)-model in a natural way. Is the converse true?

1.6.   Let Ω be the operational type of groups. We say that an Ω-term is *reduced* if it is *either* the single symbol e *or* of the form $mm \ldots mw$, where w is a string of symbols involving only variables and the operation i, and not including any substring of the form ii, ixx or xix (*except* as part of a substring $ixix$).
(i) Describe an algorithm which, given an arbitrary Ω-term t, produces a reduced term $\bar{t}$ for which $(t = \bar{t})$ is a derived equation of the theory of groups.
(ii) Show that the set of all reduced terms in a given set X of variables can be made into a group RX containing X as a subset. By considering the induced homomorphism $FX \to RX$, where FX is the free group generated by X (defined as in Theorem 1.4), show that if s and t are reduced terms for which $(s = t)$ is a derived equation, then s and t are identical.
(iii) Use (i) and (ii) to solve the word problem for groups.
[Feel free to use everything you know about group theory in answering this question.]

1.7.   (i) Let T be an algebraic theory, and suppose T contains a ternary (possibly derived) operation p for which

$$(pxyy = x) \quad \text{and} \quad (pxxy = y) \qquad\qquad (*)$$

are (possibly derived) equations of T. Let A be a T-model, and let R be a sub-T-model of $A \times A$ which contains $\{(a, a) \mid a \in A\}$ (i.e., considered as a binary relation on A, R is reflexive). Show that R is also symmetric and transitive.
(ii) Conversely, if T is an algebraic theory such that every reflexive submodel of the square of a T-model is also symmetric, show that T contains a ternary operation satisfying (*). [Hint: let F be the free T-model generated by $\{x, y\}$, and consider the sub-T-model of $F \times F$ generated by $\{(x, x), (x, y), (y, y)\}$.]

(iii) Give an example of an operation p satisfying (∗) when T is the theory of groups, but show that there is no such operation in the theory of semigroups (i.e. the theory obtained from that of groups by deleting the operation i and the equation in which i occurs).

1.8.   (i) Let $\Omega = \{e, m\}$ with $\alpha(e) = 0$, $\alpha(m) = 2$, and let E consist of the two equations $(mex = x)$ and $(mxe = x)$. Suppose a set A has two (Ω, E)-model structures (e_1, m_1) and (e_2, m_2) such that the operations of the second structure are Ω-homomorphisms $1 \to A$ and $A \times A \to A$ for the first structure. Show that A satisfies the equations $(e_1 = e_2)$ and $(m_1 m_2 xzm_2 yt = m_2 m_1 xym_1 zt)$, and deduce that $m_1 = m_2$ and that m_1 is commutative and associative.

(ii) Ask an algebraic topologist to explain what this has to do with the result that the fundamental group of a topological group is abelian.

1.9.   Let $2 = \{0, 1\}$ with its usual Boolean algebra structure, and let n be a natural number. Show that every function $2^n \to 2$ is (the interpretation of) an n-ary derived operation of the theory of Boolean algebras. [Hint: use induction on n.] Deduce that the free Boolean algebra on n generators has 2^{2^n} elements.

1.10. Let B be the theory of Boolean algebras, and let $\downarrow$ be the (derived) binary operation of B defined by

$$(x \downarrow y) = \neg(x \wedge y).$$

Show that the subtheory B_0 of B generated by $\downarrow$ (i.e. the set of all operations derivable from $\downarrow$) contains all of B except the two constants. Show also that no single operation can generate the whole of B; and that B cannot be generated by either $\wedge$ or $\vee$ plus one other operation.

2

Propositional calculus

The names which we gave to the operations of the theory of Boolean algebras, at the end of the last chapter, indicated that we were thinking of the variables to which they were applied as 'propositions' to be combined and manipulated according to rules of logic. In this chapter we make that idea more explicit; one symptom of this is that we shall change our notation and terminology slightly. We shall use the letters p, q, r rather than x, y, z to denote variables, and call them *primitive propositions*; by a *compound proposition* we shall mean a $\{\perp, \Rightarrow\}$-term in some set P of primitive propositions (or else the element of the free Boolean algebra FP which it represents). We shall refer to the elements 0, 1 of the fundamental Boolean algebra as *truth-values*; by a *valuation* of P, we mean a function $v: P \to 2$ (which of course corresponds to a Boolean homomorphism $\bar{v}: FP \to 2$).

The notion of tautology, which we introduced in the previous chapter, is a special case of that of semantic entailment. If S is a set of propositions and t a single proposition, we say that S *semantically entails* t (and write $S \models t$) if, for every valuation v (of the primitive propositions involved in $S \cup \{t\}$) such that $\bar{v}(s) = 1$ for all $s \in S$, we also have $\bar{v}(t) = 1$; i.e. 't is true whenever S is'. Clearly, t is a tautology iff $\varnothing \models t$, which we abbreviate to $\models t$.

An important example of a semantic entailment is

$$\{p, (p \Rightarrow q)\} \models q.$$

Note that if $S \models t$ is a semantic entailment, then so is any substitution instance of it, i.e. anything obtained by substituting terms for the primitive propositions appearing in $S \cup \{t\}$. Thus for example $\{p, (p \Rightarrow (q \Rightarrow p))\} \models (q \Rightarrow p)$ is a semantic entailment. Note also that if

$S \cup \{s\} \models t$ and s is a tautology, then $S \models t$, since the presence of s amongst the premises of the entailment does not affect anything.

The last few remarks represent the beginning of a notion of *proof* for propositions, i.e. of the idea that all valid semantic entailments should be obtainable by a deductive process from certain basic ones. We now set up a formal system of proofs. We begin by adopting as *axioms* all substitution instances of the three propositions

(a) $(p \Rightarrow (q \Rightarrow p))$
(b) $((p \Rightarrow (q \Rightarrow r)) \Rightarrow ((p \Rightarrow q) \Rightarrow (p \Rightarrow r)))$
(c) $(\neg \neg p \Rightarrow p)$

which we observed to be tautologies at the end of the last chapter. And we have one *rule of inference*, which is known by the Latin name *modus ponens*: from p and $(p \Rightarrow q)$, we may infer q. (Again, we allow substitution instances of this rule.)

We now define the relation of *syntactic entailment* inductively: we say a set S (whose members we shall call *premisses* of the entailment) *syntactically entails* a proposition t (written $S \vdash t$) if

(i) t is an axiom,
(ii) $t \in S$, or
(iii) for some s, we have $S \vdash s$ and $S \vdash (s \Rightarrow t)$.

In conjunction with this definition, it is convenient to introduce the notion of a *deduction from S*: a deduction of t from S is a finite sequence of propositions such that each member of the sequence is either an axiom or a member of S or is obtained via modus ponens from two earlier members of the sequence, and the last member of the sequence is t. Clearly, $S \vdash t$ iff there exists a deduction of t from S. A deduction from the empty set of premisses is also called a *proof*; if $\varnothing \vdash t$ (which we abbreviate to $\vdash t$) we say t is a *theorem* (of the propositional calculus).

To illustrate the notion of deduction, we give two examples.

Example 2.1. The following is a proof of the theorem $(p \Rightarrow p)$:

$(p \Rightarrow ((p \Rightarrow p) \Rightarrow p))$ (instance of (a))
$((p \Rightarrow ((p \Rightarrow p) \Rightarrow p)) \Rightarrow ((p \Rightarrow (p \Rightarrow p)) \Rightarrow (p \Rightarrow p)))$
                                                                 (instance of (b))
$((p \Rightarrow (p \Rightarrow p)) \Rightarrow (p \Rightarrow p))$ (modus ponens)

$(p \Rightarrow (p \Rightarrow p))$ (instance of (a))
$(p \Rightarrow p)$ (modus ponens).

Example 2.2. We deduce $(p \Rightarrow r)$ from $\{(p \Rightarrow q), (q \Rightarrow r)\}$:

$(q \Rightarrow r)$	(premiss)
$((q \Rightarrow r) \Rightarrow (p \Rightarrow (q \Rightarrow r)))$	(instance of (a))
$(p \Rightarrow (q \Rightarrow r))$	(modus ponens)
$((p \Rightarrow (q \Rightarrow r)) \Rightarrow ((p \Rightarrow q) \Rightarrow (p \Rightarrow r)))$	(instance of (b))
$((p \Rightarrow q) \Rightarrow (p \Rightarrow r))$	(modus ponens)
$(p \Rightarrow q)$	(premiss)
$(p \Rightarrow r)$	(modus ponens).

We now embark on the proof of the Completeness Theorem for the Propositional Calculus, which in the notation we have developed is the assertion that the relations $\vdash$ and $\vDash$ coincide. In one direction, the implication is straightforward:

Proposition 2.3 (the Soundness Theorem). If $S \vdash t$, then $S \vDash t$; in particular, every theorem is a tautology.

Proof. We have to show that the set $\{t \mid S \vDash t\}$ satisfies the closure conditions in the definition of $\vdash$. It certainly contains the axioms, since they are tautologies; and it trivially contains the members of S. Finally, if $S \vDash s$ and $S \vDash (s \Rightarrow t)$, then since $\{s, (s \Rightarrow t)\} \vDash t$ we have $S \vDash t$. $\square$

For the converse direction, an important tool is the following:

Theorem 2.4 (the Deduction Theorem). $S \vdash (s \Rightarrow t)$ iff $S \cup \{s\} \vdash t$.

Proof. If $S \vdash (s \Rightarrow t)$, we may obtain a deduction of t from $S \cup \{s\}$ by writing down a deduction of $(s \Rightarrow t)$ from S and adding on the two propositions s (premiss) and t (modus ponens). Conversely, let $(t_1, t_2, \ldots, t_n = t)$ be a deduction from $S \cup \{s\}$. We shall show by induction on i that $S \vdash (s \Rightarrow t_i)$ for each i.

If t_i is an axiom, or a member of S, we write down the deduction

t_i	(axiom or premiss)
$(t_i \Rightarrow (s \Rightarrow t_i))$	(instance of (a))
$(s \Rightarrow t_i)$	(modus ponens).

If t_i is s itself, we write down the proof of $(s \Rightarrow s)$ given in Example 2.1. There remains the case when we have $j, k < i$ such that t_k is $(t_j \Rightarrow t_i)$. In this case, by inductive hypothesis, we have deductions of $(s \Rightarrow t_j)$ and of $(s \Rightarrow (t_j \Rightarrow t_i))$ from S. So we write both of these down, and then add

$((s \Rightarrow (t_j \Rightarrow t_i)) \Rightarrow ((s \Rightarrow t_j) \Rightarrow (s \Rightarrow t_i)))$ (instance of (b))

$((s \Rightarrow t_j) \Rightarrow (s \Rightarrow t_i))$ (modus ponens)

$(s \Rightarrow t_i)$ (modus ponens). $\square$

The Deduction Theorem tells us that the abstract symbol $\Rightarrow$ really does behave like implication in formal proofs. We can often use it to simplify derivations: for example, to establish $\{(p \Rightarrow q), (q \Rightarrow r)\} \vdash (p \Rightarrow r)$, instead of writing down the deduction in Example 2.2, it suffices to show $\{(p \Rightarrow q), (q \Rightarrow r), p\} \vdash r$. And for this we have only to write down the three premisses and apply modus ponens twice.

The Deduction Theorem also allows us to reduce the general Completeness Theorem to a particular case, sometimes known as the Adequacy Theorem. We say that a set S of propositions is *inconsistent* if $S \vdash \perp$.

Lemma 2.5 (the Adequacy Theorem). If $S \not\models \perp$, then S is inconsistent.

Note that '$S \not\models \perp$' simply means that there is no valuation of the primitive propositions involved in S for which every member of S has truth-value 1. Before giving the proof of Lemma 2.5, we use it to obtain

Theorem 2.6 (the Completeness Theorem). $S \models t$ iff $S \vdash t$.

Proof. One direction is Proposition 2.3. Conversely, suppose $S \models t$. Then since $\{t, \neg t\} \models \perp$, we clearly have $S \cup \{\neg t\} \models \perp$. By Lemma 2.5, we have $S \cup \{\neg t\} \vdash \perp$. By Theorem 2.4, we have $S \vdash (\neg t \Rightarrow \perp)$, i.e. $S \vdash \neg \neg t$. But we may take a deduction of $\neg \neg t$ from S, and adjoin

$(\neg \neg t \Rightarrow t)$ (instance of (c))

t (modus ponens)

to obtain a deduction of t from S. $\square$

Note that the above argument is the only place so far where we have made use of axiom (c).

Proof of Lemma 2.5. We shall give the proof only in the case when the set P of primitive propositions (and hence also the set of compound propositions) is countable; the general case is similar, but requires the technical machinery of Zorn's Lemma (see Exercise 2.5, and also Chapter 7 below).

Suppose S is consistent. Then we have to show that there is a valuation v of P with $\bar{v}(t) = 1$ for all $t \in S$. Our technique for doing this will be to enlarge S as much as possible while keeping it consistent. Note first that, for any t, either $S \cup \{t\}$ or $S \cup \{\neg t\}$ is consistent; for if $S \cup \{t\}$ is inconsistent, then by the Deduction Theorem we have $S \vdash \neg t$, and so $S \cup \{\neg t\}$ has the same syntactic consequences as S itself.

We now enumerate the compound propositions, and go through them one by one: at each stage, if we can consistently adjoin the proposition under consideration to the set we have accumulated so far we do so, and otherwise we add its negation. At the end of this process, we have a set $S' \supseteq S$ which is consistent (since a deduction of $\perp$ from it would involve only finitely many of its members, and so would imply the inconsistency of the set constructed after finitely many steps) and such that, for every t, at least one of t and $\neg t$ (in fact exactly one, because $\{t, \neg t\}$ is inconsistent) belongs to S'.

We now define a function $\bar{v}$ by $\bar{v}(t) = 1$ if $t \in S'$, $\bar{v}(t) = 0$ otherwise. We claim that $\bar{v}$ is a $\{\perp, \Rightarrow\}$-homomorphism, and hence is the unique homomorphism extending its restriction to the set of primitive propositions; this will clearly suffice to prove the Lemma, since $\bar{v}(t) = 1$ for all $t \in S$. Clearly $\bar{v}(\perp) = 0$, since $\perp$ cannot belong to any consistent set. To show that $\bar{v}(s \Rightarrow t) = (\bar{v}(s) \Rightarrow_2 \bar{v}(t))$, we shall consider three cases:

(i) $\bar{v}(t) = 1$, i.e. $t \in S'$. Then we cannot have $\neg(s \Rightarrow t) \in S'$, since $t \vdash (s \Rightarrow t)$ and S' is consistent. So $(s \Rightarrow t) \in S'$, i.e. $\bar{v}(s \Rightarrow t) = 1 = (\bar{v}(s) \Rightarrow_2 \bar{v}(t))$.

(ii) $\bar{v}(s) = 0$. In this case we have $\neg s \in S'$, and since $\neg s \vdash (s \Rightarrow t)$ (see Exercise 2.2), we deduce as in (i) that we must have $\bar{v}(s \Rightarrow t) = 1 = (\bar{v}(s) \Rightarrow_2 \bar{v}(t))$.

(iii) $\bar{v}(s) = 1$, $\bar{v}(t) = 0$. In this case we have $s \in S'$ and $t \notin S'$, so since $\{s, (s \Rightarrow t)\} \vdash t$ we must have $(s \Rightarrow t) \notin S'$. Hence $\bar{v}(s \Rightarrow t) = 0 = (\bar{v}(s) \Rightarrow_2 \bar{v}(t))$. $\square$

We conclude this chapter by mentioning two applications of the Completeness Theorem. Each of them takes a property which is obvious for one of the relations $\vdash, \models$, and applies it to the other one.

Corollary 2.7 (the Compactness Theorem). If $S \models t$, then there is a finite subset $S' \subseteq S$ such that $S' \models t$.

Proof. This is obvious for $\vdash$, since a deduction of t from S can only involve finitely many of the premisses. $\square$

The name 'Compactness Theorem' is more than a fanciful analogy; this result really is tantamount to the assertion that a certain topological space is compact (see Exercise 2.6).

Corollary 2.8 (the Decidability Theorem). There is an algorithm which, given a finite set S of propositions and a proposition t, determines whether or not $S \vdash t$.

Proof. To determine whether or not $S \models t$, we have only to compute the truth-values of the members of S and of t for each of the 2^n valuations of the primitive propositions involved in $S \cup \{t\}$ (where n is the number of such propositions). $\square$

Observe that even without the Completeness Theorem it would be easy to construct an algorithm which, given S, would enumerate all possible syntactic consequences of S; but if $S \not\vdash t$, there is no way in which such an algorithm could produce this information in a finite time.

Exercises

2.1. Which of the following expressions are tautologies?
   (i) $(((p \Rightarrow q) \Rightarrow p) \Rightarrow p)$
   (ii) $(((p \vee q) \wedge \neg p) \Rightarrow q)$
   (iii) $(p \Rightarrow (\neg p \vee q))$
   (iv) $(((p \wedge \neg q) \Rightarrow r) \Rightarrow q)$
   (v) $(((p \Rightarrow q) \wedge (r \Rightarrow s)) \Rightarrow ((p \wedge r) \Rightarrow (q \vee s)))$
   (vi) $(((p \Rightarrow q) \vee (r \Rightarrow s)) \Rightarrow ((p \wedge r) \Rightarrow (q \vee s)))$.

2.2. Write down a proof of $(\bot \Rightarrow q)$ in the propositional calculus. Hence obtain a proof of $(\neg p \Rightarrow (p \Rightarrow q))$.

2.3. Use the Deduction Theorem to show that the converse of axiom (c) (i.e. the proposition $(p \Rightarrow \neg \neg p)$) is a theorem of the propositional calculus.

2.4. (For masochists only.) Write down a deduction of $(p \wedge q)$ (i.e. of $((p \Rightarrow (q \Rightarrow \perp)) \Rightarrow \perp))$ from $\{p, q\}$. [Hint: first write down a deduction of $\perp$ from $\{p, q, (p \Rightarrow (q \Rightarrow \perp))\}$, and then use the method described in the proof of the Deduction Theorem to convert it into the required deduction.]

2.5. (This question is intended for those who already have some acquaintance with Zorn's Lemma.) Let T be the set of all compound propositions in a (not necessarily countable) set P of primitive propositions.
(i) If $\{C_i \mid i \in I\}$ is a family of consistent subsets of T which is totally ordered by inclusion (i.e. for each $i, j \in I$ we have either $C_i \subseteq C_j$ or $C_j \subseteq C_i$), prove that $\bigcup_{i \in I} C_i$ is consistent.
(ii) Using Zorn's Lemma, deduce that each consistent subset of T is contained in a maximal consistent subset.
(iii) If M is a maximal consistent subset of T, show that for each $t \in T$ we have either $t \in M$ or $\neg t \in M$.
(iv) Prove Lemma 2.5 without the assumption that the set P of primitive propositions is countable.

2.6. Let P and T be as in the last question, and let V be the set of all valuations of P. For each $t \in T$, define
$$U(t) = \{v \in V \mid \bar{v}(t) = 1\}.$$
(i) Show that $U(t_1) \cap U(t_2) = U(t_1 \wedge t_2)$, and deduce that $\{U(t) \mid t \in T\}$ is a base for a topology on V.
(ii) If S is a subset of T, show that $S \models \perp$ iff $\{U(\neg t) \mid t \in S\}$ covers V.
(iii) Deduce that the Compactness Theorem (2.7) is equivalent to the assertion that the space V is compact.

3

First-order theories

So far, we have considered equational logic (where we could formulate assertions (equations) about the mathematical structures we were discussing, but didn't have any way of combining them) and propositional logic (where we studied logical combinations of assertions, but didn't attach any intrinsic meaning to the primitive propositions which were being combined). The obvious next step is to put these two ideas together. Actually, we shall miss out the next step in a traditional development of logic, the study of the first-order predicate calculus, and pass straight on to the predicate calculus with equality; predicate calculus without equality is generally similar but less interesting.

We wish to define a notion of formal language which will be adequate for handling a wide variety of mathematical structures. In addition to the features we have met so far (primitive operations, equations between terms, propositional connectives), there will be two new ones: primitive predicates and first-order quantifiers. An (n-ary) primitive predicate (ϕ, say) is something whose intended interpretation in a structure A is a subset of A^n; thus if $a_1, \ldots, a_n$ are elements of A, $\phi(a_1, \ldots, a_n)$ will be the assertion that the n-tuple $(a_1, \ldots, a_n)$ belongs to the subset which interprets ϕ. A quantifier is a unary operation on assertions of the type represented by the English phrases 'for all $x \ldots$' or 'there exists x such that $\ldots$'; the qualification 'first-order' means that the 'x' being quantified is understood to range over elements of the structure A (rather than, for example, arbitrary subsets of A).

Our language $\mathcal{L}$ will be specified by two things: an *operational type* (Ω, α), which means exactly what it did in Chapter 1, and a *predicate type* (Π, α), whose formal definition is the same as that of an operational type but whose intended interpretation is as a set of predicates with specified arities. The language itself then consists of the following things:

(1) *Variables* $x_1, x_2, x_3, \ldots$ [We assume we are given a countably infinite supply of these; of course, at any moment we shall use only a finite number of them, but it is important that the supply should never be exhausted.]

(2) *Terms* which are defined inductively, as in Chapter 1, by the following:

   (a) Every variable is a term.
   (b) If $\omega \in \Omega$, $\alpha(\omega) = n$ and $t_1, \ldots, t_n$ are terms, then $\omega t_1 \ldots t_n$ is a term.
   (c) That's all.

(3) *Atomic formulae* which are of two kinds:

   (a) If s and t are terms, then $(s = t)$ is an atomic formula.
   (b) If $\phi \in \Pi$, $\alpha(\phi) = n$ and $t_1, t_2, \ldots, t_n$ are terms, then $\phi(t_1, t_2, \ldots, t_n)$ is an atomic formula.
   [Note: part (a) of this definition essentially says that $=$ is a primitive predicate of arity 2. The reason why we keep it separate from the predicates in Π is that its intended interpretation in any structure A is fixed as the diagonal subset of $A \times A$, whereas the interpretations of members of Π can be specified arbitrarily.]

(4) *Formulae* which are defined by a further induction:

   (a) Atomic formulae are formulae.
   (b) $\bot$ is a formula; and if p and q are formulae so is $(p \Rightarrow q)$. [As previously, we introduce $\top$, $\neg p$, $(p \lor q)$, $(p \land q)$ and $(p \Leftrightarrow q)$ as abbreviations for certain compound formulae.]
   (c) If p is a formula and x is a variable which occurs free in p (see below), then $(\forall x)p$ is a formula. [We similarly introduce $(\exists x)p$ as an abbreviation for the formula $\neg(\forall x)\neg p$.]
   (d) That's all.

To explain clause 4(c) above, we need to introduce the notion of *free* and *bound* occurrences of variables in a formula. Informally, any occurrence of a variable in a formula in which $\forall$ does not appear is free; and in $(\forall x)p$ the quantifier 'binds' all those occurrences of x in p

which were not previously bound. Formally, we define FV(p), the set
of free variables in a term or formula p, by induction:

$$FV(x) = \{x\}$$
$$FV(\omega t_1 t_2 \ldots t_n) = \bigcup_{i=1}^{n} FV(t_i)$$
$$FV(s = t) = FV(s) \cup FV(t)$$
$$FV(\phi(t_1, t_2, \ldots, t_n)) = \bigcup_{i=1}^{n} FV(t_i)$$
$$FV(\bot) = \varnothing$$
$$FV(p \Rightarrow q) = FV(p) \cup FV(q)$$
$$FV((\forall x)p) = FV(p) - \{x\}.$$

Note that it is quite possible for the same variable to appear both free
and bound in different parts of a compound formula; but in practice it
is usually sensible to choose one's variables so that this doesn't
happen.

The intended interpretation of all this is as follows. If $\mathscr{L} = (\Omega, \Pi)$ is
a language, an $\mathscr{L}$-*structure* is a set A equipped with a function
$\omega_A: A^{\alpha(\omega)} \to A$ for each $\omega \in \Omega$, and a subset $[\phi]_A \subseteq A^{\alpha(\phi)}$ for each
$\phi \in \Pi$. We may then interpret each term t of $\mathscr{L}$ with $FV(t) \subseteq$
$\{x_1, x_2, \ldots, x_n\}$ as a function $t_A(n): A^n \to A$, just as we did in Chapter
1; and each formula p with $FV(p) \subseteq \{x_1, \ldots, x_n\}$ is interpreted as a
subset $[p]_A(n) \subseteq A^n$ (or equivalently as a function $p_A(n): A^n \to 2$,
namely the characteristic function of $[p]_A(n)$) by the following
inductive definition (in which we suppress the suffix (n) wherever
possible):

$(s = t)_A$ is the composite

$$A^n \xrightarrow{\;\;(s_A, t_A)\;\;} A^2 \xrightarrow{\;\;\delta\;\;} 2,$$

where $\delta(a, b) = 1$ if $a = b$, and 0 otherwise (equivalently,

$$[(s = t)]_A = \{(a_1, \ldots, a_n) \mid s_A(a_1, \ldots, a_n) = t_A(a_1, \ldots, a_n)\}).$$

$\phi(t_1, \ldots, t_m)_A$ is similarly

$$A^n \xrightarrow{\;\;((t_1)_A, \ldots, (t_m)_A)\;\;} A^m \xrightarrow{\;\;\phi_A\;\;} 2,$$

where ϕ_A is the characteristic function of $[\phi]_A$.

$\bot_A$ is the constant function with value 0.

$(p \Rightarrow q)_A$ is the composite

$$A^n \xrightarrow{\;\;(p_A, q_A)\;\;} 2^2 \xrightarrow{\;\;\Rightarrow_2\;\;} 2.$$

$[(\forall x_{n+1})p]_A(n)$ is the set

$$\{(a_1,\ldots,a_n)\,|\,\text{for all } a_{n+1}\in A,\ (a_1,\ldots,a_n,a_{n+1})\in[p]_A(n+1)\}.$$

We say a formula p is *satisfied* in a structure A (and write $A\models p$) if p_A is the constant function with value 1 (equivalently, $[p]_A$ is the whole of A^n). [Note: here as elsewhere, when we don't specify the arity of p_A, we think of it as being precisely the number of free variables of p – though in this case it doesn't make a lot of difference.] Note that $A\models p$ iff $A\models\bar{p}$, where $\bar{p}$ is the *universal closure* of p, i.e. the formula obtained by prefixing p with universal quantifiers for each of the free variables of p (in any order). A formula with no free variables is called a *sentence*.

By a *first-order theory* T in a language $\mathscr{L}$, we mean a set of sentences of $\mathscr{L}$ (called *axioms* of the theory); an $\mathscr{L}$-structure A is a *model* for T (written $A\models T$) if $A\models p$ for each $p\in T$. At this point we clearly need some examples to illuminate all the above definitions!

Example 3.1. We can regard any algebraic theory (Ω, E) as a first-order theory (without changing the meanings of the words 'structure' and 'model'): just take a language $\mathscr{L}$ in which Π is empty, and replace each equation in E by its universal closure.

Example 3.2. We can also express more complicated concepts by means of first-order axioms. The theory of commutative rings with 1 is algebraic, but that of fields is not (e.g. because the 1-element ring is not a field: cf. Exercise 1.4); but we can express the theory of fields by adding the first-order axioms

$$\neg(0=1)$$

and $(\forall x)(\neg(x=0)\Rightarrow(\exists y)(x\,.\,y=1))$

to those of the theory of commutative rings with 1.

Example 3.3. A *propositional language* is one in which Ω is empty, and all the predicates in Π have arity 0; a *propositional theory* is a set of axioms not involving equality (i.e. containing no subformula of the form $(s=t)$) in a propositional language. If we think of the elements of Π as primitive propositions, then (forgetting about the underlying set, which is irrelevant in this context) a structure for a propositional language is just a valuation of these propositions as considered in

Chapter 2, and a model for a theory T is a valuation which assigns truth-value 1 to all the propositions in T. Thus the notion of a model for a first-order theory includes one which was fundamental in the last chapter; and we shall see before long that the main theorems of the last chapter (completeness, compactness, etc.) have generalizations to this new context.

Example 3.4. As an example of a non-propositional theory expressed using primitive predicates rather than operations, we give the theory of projective planes. By a *projective plane* we mean a system of 'lines' and 'points' such that there is a unique line through any two points, and any two lines meet in a unique point. We express this by taking as primitive two unary predicates π ('is a point') and λ ('is a line') and a binary predicate $\in$ ('lies on'), with axioms

$$(\forall x)(\pi(x) \vee \lambda(x))$$
$$(\forall x)\neg(\pi(x) \wedge \lambda(x))$$
$$(\forall x, y)(\in(x, y) \Rightarrow (\pi(x) \wedge \lambda(y)))$$
$$(\forall x, y)((\pi(x) \wedge \pi(y) \wedge \neg(x = y)) \Rightarrow (\exists! z)(\in(x, z) \wedge \in(y, z)))$$
and $(\forall x, y)((\lambda(x) \wedge \lambda(y) \wedge \neg(x = y)) \Rightarrow (\exists! z)(\in(z, x) \wedge \in(z, y)))$,

where in the last two axioms we have introduced the abbreviation $(\exists! z)p(x, y, z)$ for $(\exists z)(p(x, y, z) \wedge (\forall t)(p(x, y, t) \Rightarrow (t = z)))$.

In fact (see Exercise 3.5) it is always possible to replace a given first-order theory by an 'equivalent' theory expressed in a language which has no primitive operations, only primitive predicates; and many accounts of first-order logic confine themselves to this case. (An intermediate position, adopted by some authors, is to allow primitive constants but not primitive operations of arity >0.) However, this restriction doesn't make life all that much easier, and it doesn't accord very well with mathematical practice – when we study groups or rings, it is surely natural to think of their structure in terms of operations rather than predicates.

[*Note for fusspots.* The operation of substituting one variable for another, which we used implicitly in Example 3.4 when we defined the unique existential quantifier ($\exists! z$), is actually more complicated than it seems, because of the distinction between free and bound variables, and the danger that a variable substituted into a formula might accidentally get bound. Formally, if p is a formula, t a term and x a variable, we define $p[t/x]$ ('p with t for x') to be the formula obtained from p on replacing each free occurrence of x by t, *provided* no free variable of t occurs bound in p; if it does, we must first replace each bound occurrence of such a variable by a bound occurrence of some new variable which didn't occur previously in either p or t. Of course, common sense is a much better guide than following rules such as the above when making substitutions.]

We now define the notion of semantic entailment for the predicate calculus, which is easily seen (using Example 3.3) to generalize that for the propositional calculus. If T is a set of *sentences* in a given first-order language and p is a sentence, we write $T \models p$ to mean that every model for T satisfies p. For formulae with free variables the notion of semantic entailment is less simple, because the free variables have to be assigned particular values in the structures under consideration; the easiest way to do this is to enrich our language $\mathscr{L}$ to a language $\mathscr{L}'$ by adding new constants corresponding to the free variables in $T \cup \{p\}$, and then to say that $T \models p$ iff $T' \models p'$, where p' is the sentence of $\mathscr{L}'$ obtained on replacing the free variables of p by the corresponding constants (and similarly for T').

Our next task is to develop a notion of proof for the predicate calculus with equality, and prove a completeness theorem saying that the corresponding notion of syntactic entailment coincides with semantic entailment as just defined. The overall strategy of the proof of the Completeness Theorem, and many of the details, are very similar to the propositional case, which we covered in the last chapter; so we shall give the proof here in slightly sketchy form, concentrating on the points where new ideas and difficulties arise.

Our deduction-system is, as before, defined by means of axioms and rules of inference. In addition to the three axioms we had in the propositional case, we shall require two new axioms to handle quantifiers, and two to handle equality; we shall also need one new rule of inference, the rule of *generalization*. Our complete list of axioms is

(a) $(p \Rightarrow (q \Rightarrow p))$
(b) $((p \Rightarrow (q \Rightarrow r)) \Rightarrow ((p \Rightarrow q) \Rightarrow (p \Rightarrow r)))$
(c) $(\neg\neg p \Rightarrow p)$
    (here p, q, r may be any formulae of $\mathscr{L}$)

(d) $((\forall x)p \Rightarrow p[t/x])$
(here p is any formula with $x \in FV(p)$, t any term whose free variables don't occur bound in p)
(e) $((\forall x)(p \Rightarrow q) \Rightarrow (p \Rightarrow (\forall x)q))$
(p, q formulae, $x \notin FV(p)$)
(f) $(\forall x)(x = x)$
(g) $(\forall x, y)((x = y) \Rightarrow (p \Rightarrow p[y/x]))$
(p any formula with $x \in FV(p)$, y not bound in p).

It is straightforward to verify that each of these axioms is a tautology, i.e. is satisfied in any $\mathscr{L}$-structure. Our rules of inference are

(MP) from p and $(p \Rightarrow q)$, we may infer q, *provided* either q has a free variable or p is a sentence, and

(Gen) from p we may infer $(\forall x)p$, *provided* x does not occur free in any premiss which has been used in the proof of p.

[The reason for the proviso about free variables in (MP) is the possibility that, if our language $\mathscr{L}$ has no constants, the underlying set of an $\mathscr{L}$-structure might be empty: if p has a free variable, then p and $(p \Rightarrow q)$ will automatically be satisfied in the empty structure, even if q is the sentence $\perp$. A more traditional way of avoiding this difficulty is to demand that the underlying set of any structure should be nonempty; once again, we have rejected this approach because it doesn't accord well with mathematical practice.]

With the above provisos about free variables, it is easy to verify that if S is a set of formulae and we have $S \models p$ and $S \models (p \Rightarrow q)$, then $S \models q$; and similarly for the rule of generalization. An easy induction then completes the proof of

Proposition 3.5 (the Soundness Theorem). If $S \vdash p$, then $S \models p$.
□

Theorem 3.6 (the Deduction Theorem). Suppose either that q has a free variable or that p is a sentence; then $S \cup \{p\} \vdash q$ iff $S \vdash (p \Rightarrow q)$.

Proof. Most of this proof can simply be copied from that of Theorem 2.4; but for the left-to-right implication we have one further case to deal with, namely when $q = (\forall x)r$ is obtained from r by an application of (Gen). In this case our inductive hypothesis yields $S \vdash (p \Rightarrow r)$, from

which we obtain $S \vdash (\forall x)(p \Rightarrow r)$ by (Gen), and then axiom (e) plus
(MP) is exactly what we need to obtain $S \vdash (p \Rightarrow (\forall x)r)$ – *unless* x
occurs free in p, in which case p can't have been used in the deduction
of r and so we actually have $S \vdash q$ (from which $S \vdash (p \Rightarrow q)$ follows by
standard methods). $\square$

Theorem 3.7 (the Completeness Theorem). $S \vdash p$ iff $S \models p$.

Proof. One direction is Proposition 3.5. For the converse, we may
immediately reduce to the case when $S \cup \{p\}$ is a set of sentences; for,
if $S \vdash p$, then it is easy to see that (in the notation introduced when
defining $\models$) $S' \vdash p'$, by substituting constants for free variables in a
deduction of p from S. Then, as in Chapter 2, the Deduction Theorem
allows us to reduce further to the case $p = \perp$; that is, we are reduced
to showing that if a first-order theory S is consistent (i.e. $S \not\vdash \perp$) then it
has a model.

As in Chapter 2, we shall do this by enlarging S to a consistent
theory which is maximal in a suitable sense – only this time we shall
have to enlarge the language as well as the set of sentences. In order to
avoid complications with Zorn's Lemma, we shall again restrict
ourselves to the case when the original language $\mathscr{L}$ is *countable* (i.e.
when $\Omega \cup \Pi$ is countable – an easy induction then shows that the set
of all formulae of $\mathscr{L}$ is countable).

We say that a theory S (in a given language $\mathscr{L}$) is *complete* if, for
every sentence p of $\mathscr{L}$, we have either $S \vdash p$ or $S \vdash \neg p$. We say S *has
witnesses* if, whenever $S \vdash (\exists x)p$ (with $\mathrm{FV}(p) = \{x\}$), there is a *closed
term t* of $\mathscr{L}$ (i.e. a term with no free variables, for example a constant)
such that $S \vdash p[t/x]$. (The term t is called a witness to the provability
of $(\exists x)p$.) We aim to show that any consistent theory S can be
extended to a consistent theory S^* (in some extension $\mathscr{L}^*$ of $\mathscr{L}$)
which is complete and has witnesses; clearly it will then suffice to
construct a model of S^*, since any such will define a model of S by
restriction.

The extension of S to a complete consistent theory can be done
without change of language; in fact we did it in the proof of Lemma
2.5. Now suppose $S \vdash (\exists x)p$, where $\mathrm{FV}(p) = \{x\}$; let $\mathscr{L}'$ be $\mathscr{L}$ plus one
new constant c, and $S' = S \cup \{p[c/x]\}$. Suppose $S' \vdash \perp$; then by the
Deduction Theorem we have $S \vdash \neg p[c/x]$. But since the constant c
does not appear anywhere in S, we may replace it by x throughout

this deduction to obtain $S \vdash \neg p$; and then by (Gen) we obtain $S \vdash (\forall x)\neg p$. But by hypothesis we have $S \vdash (\exists x)p$, i.e. $S \vdash \neg(\forall x)\neg p$; so S is inconsistent. The above argument shows that, given a consistent theory S, we may consistently adjoin a witness for a single sentence of the form $(\exists x)p$; hence we may consistently adjoin witnesses for all such sentences of $\mathscr{L}$ which are deducible from S.

Unfortunately, the two processes of completion and of adjoining witnesses tend to pull in opposite directions: when we complete a theory, we add lots of new deducible sentences of the form $(\exists x)p$, for which we may not have witnesses, and when we add witnesses we create lots of new sentences which may be neither provable nor refutable. So we have to do both processes repeatedly, as follows. Starting from our theory $S = S_0$ in a language $\mathscr{L} = \mathscr{L}_0$, we define a sequence of pairs $(\mathscr{L}_n, S_n)$ inductively: if n is even, we put $\mathscr{L}_{n+1} = \mathscr{L}_n$ and let S_{n+1} be a completion of S_n, and if n is odd we let $(\mathscr{L}_{n+1}, S_{n+1})$ be the result of adjoining witnesses for each sentence of the form $(\exists x)p$ deducible from S_n. (Note that there are only countably many such sentences, so our language remains countable at every stage.) Finally, we define $\mathscr{L}^* = \bigcup_{n=0}^{\infty} \mathscr{L}_n$ and $S^* = \bigcup_{n=0}^{\infty} S_n$; then S^* is consistent, since a deduction of $\bot$ from it would involve only finitely many of its members and so would imply the inconsistency of some S_n, and similar arguments show that S^* is complete and has witnesses.

To complete the proof of Theorem 3.7, we have to show that if a consistent theory is complete and has witnesses, then it has a model. Let S be such a theory; let C be the set of closed terms of its language, and let A be the quotient of C by the relation $\sim$, where

$$s \sim t \text{ iff } S \vdash (s = t).$$

(It follows from axioms (f) and (g) that $\sim$ is an equivalence relation on C; cf. Exercise 3.6.) We make A into an $\mathscr{L}$-structure by defining

$$\omega_A([t_1], \ldots, [t_n]) = [\omega t_1 \ldots t_n] \quad \text{for } \omega \in \Omega,\ \alpha(\omega) = n$$

and

$$\phi_A([t_1], \ldots, [t_n]) = 1 \text{ iff } S \vdash \phi(t_1, \ldots, t_n) \quad \text{for } \phi \in \Pi,\ \alpha(\phi) = n$$

(here square brackets denote $\sim$-equivalence classes; it is again easy to verify that these definitions are independent of the choice of representatives). Now a straightforward induction over the structure of p shows that, for any formula p of $\mathscr{L}$ (with $\mathrm{FV}(p) = \{x_1, \ldots, x_n\}$, say), we have

$$p_A([t_1], \ldots, [t_n]) = 1 \text{ iff } S \vdash p[t_1, \ldots, t_n/x_1, \ldots, x_n].$$

(The hypotheses that S is complete and has witnesses are both used in proving the left-to-right implication in the case when p has the form $(\forall x_{n+1})q$.) In particular, if $p \in S$, then since $S \vdash p$ we have $A \models p$; i.e. A is a model for S. $\square$

Corollary 3.8 (the Compactness Theorem). If S is a set of sentences such that every finite subset of S has a model, then S has a model.

Proof. As in the proof of Corollary 2.7, this is trivial if we replace the words 'has a model' by 'is consistent'. $\square$

However, we do not have an analogue of Corollary 2.8 for predicate logic; there is in general no algorithm for determining whether or not a given sentence p is deducible from a set of sentences S. (If there were, we could in particular solve the word problem for any algebraic theory.)

While Theorem 3.7 and Corollary 3.8 may seem to represent a highly desirable state of affairs, they are also indications of the limitations of first-order logic. For example, we have the following easy consequence of 3.8:

Proposition 3.9 (the Upward Löwenheim–Skolem Theorem). If a first-order theory T has an infinite model (i.e. one whose underlying set is infinite), then it has models of arbitrarily large cardinality.

Proof. Let T be a theory (in a language $\mathscr{L}$, say) with an infinite model A, and let I be an arbitrary set. We wish to construct a model A' of T whose cardinality is at least that of I. To do this, we define $\mathscr{L}'$ to be $\mathscr{L}$ plus a new family $(c_i \mid i \in I)$ of constants, and T' to be T plus the new axioms $\neg(c_i = c_j)$ for each $i \neq j$ in I. Now any finite subset of T' has a model, since it will mention only finitely many of the c_i and we can find distinct values for these in our T-model A. So T' has a model; but any such is just a model A' of T together with an injection $I \to A'$. $\square$

There is also a 'downward' Löwenheim–Skolem theorem, which we shall merely state without detailed proof (and only in the special case of countable languages).

Proposition 3.10 (the Downward Löwenheim–Skolem Theorem).
If a first-order theory T in a countable language has an infinite
model, then it has a countably infinite model. More
specifically, if A is any model of T, and B any countable subset
of A, then there is a countable submodel of A containing B.

The concept of submodel hasn't been defined, but it is easy to see
what it must be. To understand the idea behind the proof of
Proposition 3.10, think of what it says when T is the theory of groups:
it is then (essentially) the assertion that the subgroup of a group A
generated by a countable subset B is countable. The proof in general
is quite similar to what we would do to prove this, namely start from
B and go on adjoining elements of A whenever forced to do so by T
until the process terminates – and then show that we haven't gone
outside the realm of countable sets. (The major extra complication in
the general case is that we may have to make arbitrary choices: for
example, if T contains a sentence of the form $(\forall x)(\exists y)p$ and $b \in B$,
there may be uncountably many $a \in A$ for which $(b, a) \in [p]_A$, and we
can't afford to add them all to our submodel.) Note in passing that
the technique we used to prove 3.7 is already sufficient to prove the
first sentence of the statement of 3.10, since it produces a countable
model of any consistent theory in a countable language (we can
ensure that it is countably infinite rather than finite by using the
technique in the proof of 3.9, with I taken to be countably infinite).

A theory T is said to be *categorical* if it has only one model, up to
isomorphism (again, we haven't defined isomorphism of T-models,
but . . .). The Löwenheim–Skolem theorems tell us that a first-order
theory with an infinite model can't be categorical (although there *are*
categorical first-order theories with finite models: for example, the
theory of simple groups of order 168). In the practice of mathematics,
we are accustomed to come across axiom-systems which (in the
above sense) categorize infinite structures; so what is going on here?

A good example of such an axiom-system is *Peano's postulates* for
the natural numbers (G. Peano, 1891). These say, approximately,

   (i) 0 is a natural number. [Peano actually started at 1, but who
      cares?]
  (ii) Every natural number has a successor.

(iii) 0 is not a successor.

(iv) Distinct natural numbers have distinct successors.

(v) (The induction postulate) If p is a property of natural numbers which is true for 0, and which is true for the successor of x whenever it's true for x, then P is true for all natural numbers.

Expressed in modern terminology, postulates (i) and (ii) are not axioms; they simply define the language by saying that it contains a constant 0 and a unary operation s. Postulates (iii) and (iv) may then be expressed by first-order sentences in this language:

$$(\forall x)\neg(sx = 0)$$

and $(\forall x, y)((sx = sy) \Rightarrow (x = y))$.

The problem comes with postulate (v), which speaks (implicitly) of all possible properties of natural numbers, i.e. of all possible subsets of our intended model $\mathbb{N}$. In first-order logic we're not allowed to quantify over subsets, but only over elements; but we might try to approximate to Peano's intentions by introducing a *scheme* of axioms of the form

$$((p[0/x] \wedge (\forall x)(p \Rightarrow p[sx/x])) \Rightarrow (\forall x)p), \qquad (*)$$

one for each formula p of our language with $FV(p) = \{x\}$. Actually, we can be a little more general than this by allowing 'induction with parameters': that is, we can allow p to have additional free variables $y_1, \ldots, y_n$ and then apply $(\forall y_1, \ldots, y_n)$ to the formula $(*)$ above. If we do this, we obtain a first-order theory which we shall call 'weak Peano arithmetic'.

However, when logicians speak of 'Peano arithmetic', they generally mean an extension of this theory obtained by adjoining two binary operations a and m to the language, plus the axioms

$(\forall x)(ax0 = x)$

$(\forall x, y)(axsy = saxy)$

$(\forall x)(mx0 = 0)$

and $(\forall x, y)(mxsy = amxyx)$

which express the usual recursive definitions of addition and multiplication (and, of course, all the extra instances of $(*)$ arising from the extension of the language). First-order Peano arithmetic (which we shall abbreviate to PA) is a useful and versatile theory (though it's doubtful whether Peano himself would have recognized it), and we shall make extensive use of it in Chapter 9. But we haven't captured the full force of Peano's fifth postulate; for our language is still countable, and hence there are only countably many axioms of the form $(\forall y_1, \ldots, y_n)$ (∗) (each of which, if interpreted in a countable structure A, makes an assertion about countably many subsets of A – one for each assignment of values to the parameters $y_1, \ldots, y_n$), whereas Peano's (v) applies to each of the uncountably many subsets of $\mathbb{N}$. Thus the fact that PA has models not isomorphic to $\mathbb{N}$ is not, after all, surprising.

Similar remarks can be applied to the assertion, familiar from a first course on real analysis, that the axioms for a (Dedekind) complete ordered field categorize the real line $\mathbb{R}$. The problem is that the Dedekind completeness axiom

$$(\forall S \subseteq \mathbb{R})((S \neq \varnothing \text{ and } S \text{ has an upper bound}) \Rightarrow$$

$$(S \text{ has a least upper bound}))$$

involves a quantification over subsets of $\mathbb{R}$; once again, we can replace it by a scheme of first-order axioms in the language of ordered rings (i.e. the language of rings with an additional binary predicate $\leqslant$), but there are only countably many such axioms and the resulting theory will have a countable model.

Exercises

3.1. Formulate sets of axioms in suitable first-order languages (to be specified) for the following theories:
(i) The theory of integral domains.
(ii) The theory of algebraically closed fields of characteristic zero.
(iii) The theory of separably closed fields of characteristic 2. [A field K is said to be *separably closed* if each polynomial over K which has no common factor with its (formal) derivative has a root in K.]
(iv) The theory of local rings. [A *local ring* is a commutative ring which has a unique maximal ideal.]
(v) The theory of partially ordered sets.
(vi) The theory of partially ordered sets having greatest and least elements.
(vii) The theory of totally ordered fields.
(viii) The first-order theory of Dedekind-complete ordered fields.

(ix) The theory of simplicial complexes. [Take the underlying set to consist of vertices, with an n-ary predicate ϕ_n for each n to express the assertion that a given n-tuple of vertices spans a simplex.]

(x) The theory of groups of order 168. [Add 167 constants to the language of groups.]

(xi) The theory of simple groups of order 168.

3.2. Let $\mathscr{L}$ be the first-order language having one binary predicate ϕ_r for each real number $r \geq 0$, and let T be the $\mathscr{L}$-theory with axioms

$$(\forall x, y)(\phi_0(x, y) \Leftrightarrow (x = y)),$$
$$(\forall x, y)(\phi_r(x, y) \Rightarrow \phi_s(y, x)) \quad \text{for each } (r, s) \text{ with } r \leq s,$$

and $(\forall x, y, z)((\phi_r(x, y) \wedge \phi_s(y, z)) \Rightarrow \phi_{r+s}(x, z))$ for each (r, s).

Show that a metric space (X, d) becomes a T-model if we interpret $\phi_r(x, y)$ as '$d(x, y) \leq r$'. Is every T-model obtained from a metric space in this way?

3.3. (a) Define the notion of *substructure* of an $\mathscr{L}$-structure A.

(b) Show that if B is a substructure of A, and p is a quantifier-free formula of $\mathscr{L}$ (with n free variables, say), then

$$[p]_B = [p]_A \cap B^n.$$

[If you can't do this, your answer to (a) is probably wrong.]

(c) By considering the centre of a group, or otherwise, show that the conclusion of (b) may fail if p contains quantifiers.

3.4. (a) A first-order theory is called *universal* if its axioms all have the form $(\forall \vec{x})p$, where $\vec{x}$ is a finite (possibly empty) string of variables and p is quantifier-free. If T is a universal theory, show that every substructure of a T-model is a T-model.

(b) T is called *inductive* if its axioms have the form $(\forall \vec{x})(\exists \vec{y})p$, where p is quantifier-free. Let T be an inductive theory, and A a structure for the language of T; suppose A is the union of a family of substructures $(B_i \mid i \in I)$, where the B_i are totally ordered by inclusion and each B_i is a T-model. Show that A is a T-model.

(c) Which of the theories in Exercise 3.1 are (expressible as) universal theories? And which are inductive?

3.5. Let $\mathscr{L}$ be a first-order language having sets Ω and Π of primitive operations and predicates. Let $\mathscr{L}^*$ be the language obtained from $\mathscr{L}$ on replacing each primitive operation ω by a primitive predicate ω^* with $\alpha(\omega^*) = \alpha(\omega) + 1$. If T is a theory in $\mathscr{L}$, show how to construct a theory T^* in $\mathscr{L}^*$ which is equivalent to T in the sense that it has 'the same' models. [Hint: first write down axioms to express '$[\omega^*]$ is the graph of a function'.]

3.6.   Show that the sentences

$$(\forall x, y)((x = y) \Rightarrow (y = x))$$
and $(\forall x, y, z)((x = y) \Rightarrow ((y = z) \Rightarrow (x = z)))$

are theorems of the predicate calculus with equality.

3.7.   Show that the first-order theory T whose only axiom is

$$(\forall x)\neg(x = x)$$

is consistent iff the language in which it is written has no constants.
Show also that, if it is consistent, it is complete (provided the language
has no nullary predicate symbols) and has witnesses. [Hint: first show
that every sentence of the form $(\forall x)p$ is provable in T – you will find
Exercise 2.2 useful for this.]

3.8.   (a) Let $\mathscr{L}$ be the language of weak Peano arithmetic, and let N be an
$\mathscr{L}$-structure. Show that the following are equivalent:
(i) N is a model for the higher-order Peano postulates.
(ii) For any $\mathscr{L}$-structure A, there is a unique homomorphism $N \to A$
(i.e. N is freely generated by the empty set; cf. Theorem 1.2).
[Hint: given (i) and an $\mathscr{L}$-structure A, let S be the sub-$\mathscr{L}$-structure of
$N \times A$ generated by the empty set, and show that $S \to N \times A \to N$ is a
bijection.]
(b) Let N be an $\mathscr{L}$-structure satisfying the equivalent conditions of
part (a). By taking A to be the set N^N of all functions $N \to N$ (equipped
with suitable maps $1 \to A \to A$), show that there exist functions
$a_N : N^2 \to N$ and $m_N : N^2 \to N$ making N into a model of (first-order)
Peano arithmetic.

3.9.   Show that the sentences

$$(\forall x, y, z)(axayz = aaxyz) \quad \text{and} \quad (\forall x, y)(axy = ayx)$$

are provable in Peano arithmetic.

3.10. Let $\mathscr{L}$ be the language with one binary predicate $<$, and let T be the
$\mathscr{L}$-theory with axioms

$$(\forall x)\neg(x < x)$$
$$(\forall x, y)((x < y) \lor (y < x) \lor (x = y))$$
$$(\forall x, y, z)(((x < y) \land (y < z)) \Rightarrow (x < z))$$
$$(\forall x)(\exists y, z)((y < x) \land (x < z))$$
and $(\forall x, y)((x < y) \Rightarrow (\exists z)((x < z) \land (z < y)))$.

Show that every countable T-model is isomorphic to the ordered set $\mathbb{Q}$
of rational numbers. Is every countable model of Peano arithmetic
isomorphic to $\mathbb{N}$?

3.11. Let T be the first-order theory of Dedekind-complete ordered fields (cf. Exercise 3.1(viii)). If K is a T-model, show that K is a *real-closed field* (that is, an ordered field in which every positive element has a square root, and such that every odd-degree polynomial over K has a root in K). [In fact it can be shown that the theory of Dedekind-complete ordered fields is equivalent to the theory of real-closed fields.]

4

Recursive functions

At one or two points in the preceding chapters, we have referred to the existence or non-existence of algorithms to solve particular problems. It's easy to see how one proves that an algorithm exists: one simply constructs it, and demonstrates that it does the job one wants it to do. But how can one prove that *no* algorithm exists for a particular problem?

Clearly, in order to do this, we are going to have to be more precise than hitherto about what we mean by an algorithm. Informally, we can think of an algorithm as some calculation which a computer could be programmed to carry out; but, in order to make this precise, we need a precise definition of what we mean by a computer. In fact, our 'idealized' mathematical model of a computer will be a pretty feeble thing in comparison with most physically existing computers (we are not, on this theoretical level, interested in questions of speed or efficiency of computation, and so for simplicity we shall give our computer only the minimum of features needed for it to function at all); but in one respect it will be more powerful than the largest computer ever built – it will be able to handle *arbitrarily large* natural numbers. Of course, any physically existing machine suffers from limitations on the size of the numbers it can handle, but for our purposes it is essential that we dispense with this restriction – if we want to know whether some function $f: \mathbb{N} \to \mathbb{N}$ is algorithmically computable, it is not enough to say that we can build a machine guaranteed to give the right value of $f(n)$ for all $n < 10^{10^{10}}$, since such a machine might really be computing a different function which just happens to agree with f at these values.

The first person to formulate the idea of such a theoretical computer, and of the class of theoretically computable functions, was the British mathematician A. M. Turing, in 1936. (After the Second World War, Turing was also a pioneer in the practical development of electronic computers.) We shall not follow Turing's description of a theoretical computer exactly, but instead use a notion of 'register machine' (essentially due to M. Minsky) which has certain technical advantages over Turing's machine, and turns out to have exactly the same computing power.

A register machine, then, has a sequence of *registers* $R_1, R_2, \ldots$, each of which may at any time hold an arbitrary natural number. A *program* for the machine is defined by specifying a finite number of states $S_0, S_1, \ldots, S_n$, together with, for each $i > 0$, an instruction to be carried out whenever the machine is in state S_i. (S_0 is the *terminal state*; on reaching it, the machine 'switches itself off'.) These instructions are of two kinds:

(a) add 1 to register R_j, and move to state S_k;

(b) test whether R_j holds the number 0: if it does, move to state S_l; otherwise subtract 1 from it and move to state S_k.

We can represent these instructions diagrammatically, as follows:

$$S_i \xrightarrow{R_j+1} S_k \qquad\qquad S_i \xrightarrow{R_j-1} S_k$$
$$\searrow S_l$$

A program may be represented by putting together several such diagrams: here is a simple example which (starting from S_1, which we normally take as our *initial state*) adds the contents of R_2 to R_1, leaving R_2 empty.

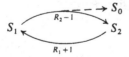

Alternatively, we may write a program as a finite sequence of triples $(j, +, k)$ or quadruples $(j, -, k, l)$ (representing the basic instructions (a) and (b) above; the ith term of the sequence is the instruction to be obeyed in state i). The following sequence represents a program which first empties R_3, then adds twice the contents of R_2 to R_1, and finishes with R_2 containing what it did initially:

$$(3, -, 1, 2)$$
$$(2, -, 3, 6)$$
$$(3, +, 4)$$
$$(1, +, 5)$$
$$(1, +, 2)$$
$$(3, -, 7, 0)$$
$$(2, +, 6).$$

[Exercise: rewrite this program as a diagram, and verify that it does what was claimed above.]

We note that any given program, since it has a finite number of instructions, will affect only finitely many of the registers of our machine; but (as with variables in the predicate calculus) it is convenient to have an unbounded supply of registers available. Normally, we think of a certain initial segment $(R_1, R_2, \ldots, R_k)$ of the registers as 'input registers', which are loaded before the start of a program with the data on which we wish to perform some computation; the remaining registers constitute the program's 'scratch pad', and are normally assumed to be set to 0 before it starts. Thus we say that a function $f: \mathbb{N}^k \to \mathbb{N}$ is *computable* by a given program P if, starting from state S_1 with the registers holding $(n_1, n_2, \ldots, n_k, 0, 0, \ldots)$, the machine eventually reaches state S_0 with $f(n_1, n_2, \ldots, n_k)$ in register R_1.

Of course, a given program may run for ever without reaching state S_0, or may reach S_0 only for certain values of the input data; so our notion of computable function includes partial functions (i.e. functions defined on subsets of $\mathbb{N}^k$) as well as globally defined ones. Here, for example, is a program which computes $n_1 - n_2$ if $n_1 \geq n_2$, and fails to terminate if $n_1 < n_2$:

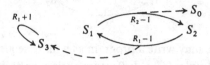

The next problem is clearly to determine which functions are computable. We begin by showing that the class of computable functions has good closure properties.

Theorem 4.1. (a) For each $i \leqslant k$, the projection function $(n_1, \ldots, n_k) \mapsto n_i$ is computable.

(b) The constant function with value 0, and the successor function $\mathbb{N} \to \mathbb{N}$, are computable.

(c) If f is a computable (partial) function of k variables and $g_1, \ldots, g_k$ are computable functions of l variables, then the function h defined by

$$h(n_1, \ldots, n_l) = f(g_1(n_1, \ldots, n_l), \ldots, g_k(n_1, \ldots, n_l))$$

is computable. [Note that if f and the g_i are not globally defined, we regard h as being defined at $(n_1, \ldots, n_l)$ iff each g_i is defined there and f is defined at the resulting k-tuple of values.]

(d) If f and g are computable functions of arities k and $k + 2$ respectively, then the $(k + 1)$-ary function h defined by

$$h(n_1, n_2, \ldots, n_k, 0) = f(n_1, n_2, \ldots, n_k)$$
$$h(n_1, \ldots, n_k, n_{k+1} + 1) = g(n_1, \ldots, n_k, n_{k+1}, h(n_1, \ldots, n_k, n_{k+1}))$$

is computable. [Again, we adopt the obvious convention about when h is undefined.]

(e) If f is a computable function of arity $k + 1$, then the k-ary function g defined by

$$g(n_1, \ldots, n_k) = n \text{ if } f(n_1, \ldots, n_k, n) = 0 \text{ and}$$
$$f(n_1, \ldots, n_k, m) > 0 \quad \text{for all } m < n$$

(and $g(n_1, \ldots, n_k)$ undefined if no such n exists) is computable.

We summarize parts (a) and (b) of this theorem by saying that the *basic functions* are computable, part (c) by saying that the class of computable functions is closed under composition, part (d) by saying that it is closed under (primitive) *recursion*, and part (e) by saying that it is closed under *minimalization*.

Proof. (a) For $i = 1$, the projection function can be computed by any program which leaves R_1 alone and is guaranteed to terminate, for example the one-line program $(2, +, 0)$. For $i > 1$, we write a program which first empties R_1 and then transfers the contents of R_i to it, thus: $(1, -, 1, 2), (i, -, 3, 0), (1, +, 2)$.

(b) These two functions can both be computed by one-line programs, namely $(1, -, 1, 0)$ and $(1, +, 0)$.

(c) We construct a program for computing h as follows. First we transfer the contents of $R_1, R_2, \ldots, R_l$ to $R_{n+1}, R_{n+2}, \ldots, R_{n+l}$ respectively, where n is chosen so large that these registers will not be disturbed in the subsequent computations. Then, for $1 \leqslant i \leqslant k$, we transfer these contents to $R_{k+1}, R_{k+2}, \ldots, R_{k+l}$ (without destroying the contents of $R_{n+1}, \ldots, R_{n+l}$), and perform our program for computing g_i (with all registers shifted k places to the right), storing the answer in R_i. Finally, we perform our program for computing f, terminating when (and if) the latter does.

(d) Here we again begin by copying our input data from $R_1, \ldots, R_{k+1}$ to a 'remote' set of registers $R_{n+1}, \ldots, R_{n+k+1}$. Next we perform our program for computing f; having reached the end of this, we obey the instruction $(n + k + 1, -, j, 0)$, where state S_j is the beginning of a 'subroutine' which transfers the contents of R_1 to R_{k+2}, then those of $R_{n+1}, \ldots, R_{n+k}$ to $R_1, \ldots, R_k$ and those of R_{n+k+2} to R_{k+1}, then performs the program for computing g and finally adds 1 to R_{n+k+2}. At the end of the subroutine we go back to the state in which we obeyed the instruction $(n + k + 1, -, j, 0)$.

(e) Once again, we first copy $R_1, \ldots, R_k$ to $R_{n+1}, \ldots, R_{n+k}$. Then we enter a subroutine in which we copy $R_{n+1}, \ldots, R_{n+k+1}$ to $R_1, \ldots, R_{k+1}$ and then perform the computation of f. At the end of the subroutine we obey $(1, -, j, j')$; from S_j we add 1 to R_{n+k+1} and return to the beginning of the subroutine, and from $S_{j'}$ we transfer the contents of R_{n+k+1} to R_1 and then terminate. $\square$

We define the class of *recursive functions* to be the smallest class of (partial) functions from powers of $\mathbb{N}$ to $\mathbb{N}$ having the closure properties of Theorem 4.1; thus a function is recursive iff it can be constructed from the basic functions by a finite number of applications of composition, primitive recursion and minimalization. If f can be constructed without using minimalization then we call it *primitive recursive*. Note that if the input data in Theorem 4.1(c) or (d) consists of globally defined functions, then so does the output; thus every primitive recursive function is globally defined. (However, not every globally defined recursive function is primitive recursive, as we shall see later.) As an example, the recursion equations

$$n_1 + 0 = n_1$$
$$n_1 + (n_2 + 1) = (n_1 + n_2) + 1$$

and

$$n_1 . 0 = 0$$
$$n_1 . (n_2 + 1) = (n_1 . n_2) + n_1$$

tell us successively that the addition and multiplication functions $\mathbb{N} \times \mathbb{N} \rightarrow \mathbb{N}$ are primitive recursive. Hence (the interpretation in $\mathbb{N}$ of) every term of the language of PA is a primitive recursive function. (We shall return to this point later – see Theorem 4.13.)

Theorem 4.1 tells us in particular that every recursive function is computable. In the other direction, we have

Theorem 4.2. Every computable function is recursive.

Proof. Let P be a program for computing a (partial) k-ary function f. We define an auxiliary function g, of arity $k + 2$, as follows:

$g(n_1, \ldots, n_k, 0, t)$ is the number of the state reached after P has been running for t steps, given that it started in state S_1 with the registers set to $(n_1, \ldots, n_k, 0, 0, \ldots)$ (it is understood that if P terminates in fewer than t steps, then it remains in state 0 after it has terminated).

For $i > 0$, $g(n_1, \ldots, n_k, i, t)$ is the contents of R_i after P has been running for t steps (with the same convention about what happens after P has terminated).

Clearly, g is globally defined. It is not immediately obvious that g is recursive (in fact it is primitive recursive), since the value of $g(n_1, \ldots, n_k, i, t + 1)$ depends on that of $g(n_1, \ldots, n_k, i', t)$ for values of i' other than i; it thus appears that we need a double recursion to compute it, rather than a primitive recursion as in Theorem 4.1(d). However, there are only finitely many values of i for which $g(n_1, \ldots, n_k, i, t)$ can ever be nonzero; and we can 'code' the finite sequence $(g_0, g_1, \ldots, g_r)$ of its values for $0 \leqslant i \leqslant r$ (and some fixed $n_1, \ldots, n_k$ and t) by the single integer $2^{g_0} 3^{g_1} \ldots p_r^{g_r}$, where p_i denotes the $(i + 1)$th prime number. It is clear that the 'coding' function which sends $(g_0, \ldots, g_r)$ to this product is primitive recursive; it is less clear that the 'decoding' functions $n \mapsto (n)_i$ are so, where $(n)_i$ denotes the exponent of the largest power of p_i dividing n, but see Exercise 4.3

below. Given this, it is easy to show that the function which sends $(n_1, \ldots, n_k, t)$ to the product defined above is primitive recursive, and hence that g is.

Now we observe that $f(n_1, \ldots, n_k) = g(n_1, \ldots, n_k, 1, h(n_1, \ldots, n_k))$ where $h(n_1, \ldots, n_k)$ is the least t such that $g(n_1, \ldots, n_k, 0, t)$ is 0, if this exists (and the $=$ sign is interpreted as meaning that the expression on one side is defined iff the other is and then they are equal). So f is recursive. $\square$

Theorems 4.1 and 4.2 together constitute one instance of *Church's Thesis* (named after A. Church, who introduced the class of recursive functions – via a different definition, which we shall not explore here – independently of, and at the same time as, Turing); the thesis asserts that the recursive functions coincide with the computable functions for *any* reasonable 'abstract theory of computation'. As a general assertion, Church's Thesis is obviously not capable of formal proof – it is always possible that somebody will devise a new theory of computation which is capable of computing more functions than our register machine – but it has been verified for every such theory which has been proposed so far, and is generally accepted as true by those who work with recursive functions. Henceforth we shall assume it in the following informal sense: if we wish to show that a given function is recursive, we shall regard it as sufficient to give an informal description of an algorithm for computing it, without verifying explicitly that this description can be translated into a register machine program.

Clearly, not every function $\mathbb{N}^k \to \mathbb{N}$ is recursive, since there are uncountably many such functions, but only countably many programs for our register machine. We can code programs by natural numbers, as follows: first we replace the triples $(j, +, k)$ and quadruples $(j, -, k, l)$ which form the individual instructions by numbers $2^j \cdot 5^k$ and $2^j \cdot 3 \cdot 5^k \cdot 7^l$ respectively, and then we replace the finite sequence of numbers $i_1, i_2, \ldots, i_r$ which correspond to the instructions in our program by the single number $2^{i_1} \cdot 3^{i_2} \ldots p_{r-1}^{i_r}$. We write P_n for the program coded in this way by the natural number n. Of course, not every natural number will code a program in this way – for example, if n codes a program then each of the numbers $(n)_r$, defined as in the proof of Theorem 4.2, is divisible only by primes $\leqslant 7$

– but it is easy to construct an algorithm which determines whether n codes a program, i.e. the function f defined by

$$f(n) = 1 \quad \text{if } n \text{ codes a program}$$
$$= 0 \quad \text{otherwise}$$

is recursive. We say that a set of natural numbers (or more generally a subset of $\mathbb{N}^k$ for some k) is a *recursive set* if its characteristic function is a recursive function, as here.

Of course, a program does not uniquely specify the function which it computes, since if P computes a k-ary function f it also computes the $(k-1)$-ary function f' defined by

$$f'(n_1, \ldots, n_{k-1}) = f(n_1, \ldots, n_{k-1}, 0).$$

But f is uniquely specified by giving its arity together with a program for computing it; we shall write $f_{n,k}$ for the k-ary (partial) function computed by P_n (if P_n exists). Adapting Cantor's diagonal argument, we can now give an explicit definition of a function $\mathbb{N} \to \mathbb{N}$ which is not recursive: if we define

$$g(n) = f_{n,1}(n) + 1 \quad \text{if } (f_{n,1} \text{ exists and}) \ f_{n,1}(n) \text{ is defined}$$
$$= 0 \qquad\qquad \text{otherwise,}$$

then clearly $g \neq f_{n,1}$ for any n.

At first sight, this is rather surprising, since the definition of g just given appears to yield an algorithm for computing its value at any natural number n. However, this is not so, since the 'otherwise' clause of the definition includes cases where $f_{n,1}$ exists but the computation of $f_{n,1}(n)$ does not terminate; in such cases we shall not be able to deduce the value of $g(n)$ from any *finite* computation. If we modified the definition of g by leaving it undefined at these values of n, then it would be recursive: this follows easily from

Theorem 4.3. There exists a recursive function u (of three variables) such that

$$u(n, k, m) = r \quad \text{if } n \text{ codes a program, } m \text{ codes a } k\text{-tuple}$$
$$((m)_1, \ldots, (m)_k) \text{ of natural numbers and}$$
$$f_{n,k}((m)_1, \ldots, (m)_k) \text{ (is defined and) equals } r,$$

and $u(n, k, m)$ is undefined otherwise.

Proof. We give an informal description of an algorithm for computing u. First we decode n and check whether it is a code for a program (if not, we enter some non-terminating loop), and then we decode m as a k-tuple, if possible. Then we 'simulate' the operation of P_n on the input data $((m)_1, \ldots, (m)_k, 0, 0, \ldots)$, by computing the auxiliary function $g((m)_1, \ldots, (m)_k, i, t)$ defined in the proof of Theorem 4.2, for the relevant range of values of i and for successive values of t. (It is clear that this can be done algorithmically from the input data (n, k, m), in such a way that the computation of each successive value can be guaranteed to terminate in a finite time.) If we ever reach a value of t for which $g((m)_1, \ldots, (m)_k, 0, t) = 0$, we compute the corresponding $g((m)_1, \ldots, (m)_k, 1, t)$ and then terminate; otherwise we go on computing for ever. □

A program for computing a function like u of Theorem 4.3 is said to be *universal*, since it is capable of simulating the action of *any* program that we can write for our register machine. It is possible to give explicit examples of universal programs, but it would take too long to do so here.

Next we consider a close relative of the notion of a recursive set.

Lemma 4.4. For a subset E of $\mathbb{N}$, the following are equivalent:
    (a) E is the set of values taken by some recursive function (possibly of several variables).
    (b) E is the domain of definition of some (partial) recursive function of 1 variable.
    (c) The function ϕ_E defined by $\phi_E(n) = 0$ if $n \in E$, undefined otherwise, is recursive.
    (d) The function ψ_E defined by $\psi_E(n) = n$ if $n \in E$, undefined otherwise, is recursive.
    Moreover, E is a recursive set iff E and $\mathbb{N} - E$ both satisfy the above conditions.

Proof. (b) $\Rightarrow$ (c): Given a program for computing some function f with domain E, we may modify it to compute ϕ_E by adding an instruction which resets R_1 to 0 before the program terminates.

(c) $\Rightarrow$ (d): Similarly, a program for computing ϕ_E may easily be modified to compute ψ_E; and (d) $\Rightarrow$ (a) is immediate.

(a) $\Rightarrow$ (b): Given a program P which computes a function f whose range of values is E, we set up a new program Q which, given an input n, systematically computes P for all possible values of the input to P

(doing finitely many steps at a time from each computation, in such a way that it is bound to reach the end of each terminating computation in a finite time). Each time this computation produces a value of f, Q compares it with the input n; if the two are equal, Q terminates, otherwise it continues with the computation of values of f.

For the last part, if the characteristic function χ_E of E is computable then so are ϕ_E and ϕ_{N-E}. Conversely, if ϕ_E and ϕ_{N-E} are both computable we may write a program to run both computations simultaneously, and produce the output 1 or 0 according to which one finishes first. □

A set satisfying the conditions (a)–(d) of Lemma 4.4 is called *recursively enumerable*. Not every such set is recursive:

Example 4.5. The set $\{n \in \mathbb{N} \mid u(n, 1, 3^n)$ is defined$\}$ (where u is the function defined in Theorem 4.3) is recursively enumerable but not recursive. The first statement is obvious from the definition of the set; but if the set were recursive, then it would be easy to write a program for computing the function g defined before Theorem 4.3, since $g(n) = u(n, 1, 3^n) + 1$ whenever the latter is defined, and $g(n) = 0$ otherwise.

Example 4.6. Similarly, we may show that the set

$$E = \{n \in \mathbb{N} \mid f_{n,1} \text{ is globally defined}\}$$

is not even recursively enumerable. For, if it were, then (cf. Exercise 4.4) we could find a globally defined recursive function (h, say) whose range of values is E, and then define

$$g(n) = f_{h(n),1}(n) + 1 = u(h(n), 1, 3^n) + 1.$$

Then g is recursive, and globally defined, but $g \neq f_{m,1}$ for any $m \in E$.

The fact that the set of Example 4.5 (and hence also the set $\{(n, k, m) \mid u(n, k, m)$ is defined$\}$) fails to be recursive is generally expressed by saying that the *halting problem* for register machines is undecidable: that is, there is no algorithm which, given codes for a program and for an input to it, determines whether the program will terminate starting from the given input. The undecidability of other mathematical problems, such as those to which we alluded in Chapters 1 and 3, is generally proved by showing that, given an algorithm for solving the problem under discussion, we could construct one to solve the halting problem. (See Example 4.11 below.)

Let f be a recursive function of $(k + l)$ variables, say $f = f_{n,k+l}$. If we fix the values of the first k variables at $a_1, a_2, \ldots, a_k$, then we have a recursive function g of l variables defined by

$$g(n_1, \ldots, n_l) = f(a_1, \ldots, a_k, n_1, \ldots, n_l).$$

Clearly, we can write $g = f_{m,l}$ for some m (in fact for infinitely many m, since there are infinitely many ways in which we can introduce inessential modifications into a program for computing g); however, the question arises whether we can choose a suitable m in some algorithmic way, given a code n for a program which computes f and the 'parameter' values $a_1, \ldots, a_k$. The answer, pretty clearly, is yes: given a program for computing f, we simply prefix it by a program which transfers the contents of $R_1, \ldots, R_l$ to $R_{k+1}, \ldots, R_{k+l}$ and then loads $R_1, \ldots, R_k$ with the given parameter values. Since this process is algorithmic, it corresponds by Church's Thesis to a recursive function of the $k + 1$ variables $n, a_1, \ldots, a_k$. Thus we have

Theorem 4.7. For each pair (k, l) of positive integers, there is a globally defined recursive function $t_{k,l}$ of $k + 1$ variables such that

$$f_{n,k+l}(a_1, \ldots, a_k, b_1, \ldots, b_l) = f_{t_{k,l}(n,a_1,\ldots,a_k),l}(b_1, \ldots, b_l)$$

for all $n, a_1, \ldots, a_k, b_1, \ldots, b_l$ (in the sense that one side is defined iff the other is and then they are equal).   □

Theorem 4.7 (like a good deal of recursion theory) is due to S. C. Kleene; it is commonly called the 's–m–n theorem', a meaningless name derived from the notation originally used by Kleene (and copied by most subsequent writers). In conjunction with the existence of a universal program (4.3), it is a very powerful tool for showing that certain functions are recursive. For example,

Lemma 4.8. For each k, there is a recursive function r_k of two variables which 'codes primitive recursion', in the sense that

$$f_{r_k(n,m),k+1}(x_1, \ldots, x_k, 0) = f_{n,k}(x_1, \ldots, x_k)$$

and

$$f_{r_k(n,m),k+1}(x_1,\ldots,x_k,x_{k+1}+1)$$
$$= f_{m,k+2}(x_1,\ldots,x_k,x_{k+1},f_{r_k(n,m),k+1}(x_1,\ldots,x_k,x_{k+1})).$$

Proof. Consider the $(k+3)$-ary function g defined by

$$g(n,m,x_1,\ldots,x_k,x_{k+1}) = f(x_1,\ldots,x_k,x_{k+1})$$

where f is the $(k+1)$-ary function defined (as above) by recursion from $f_{n,k}$ and $f_{m,k+2}$. It is clear that g is recursive, since it may be defined by primitive recursion using u:

$$g(n,m,x_1,\ldots,x_k,0) = u(n,k,3^{x_1}5^{x_2}\ldots p_k^{x_k}),$$

with a similar expression for $g(n,m,x_1,\ldots,x_k,x_{k+1}+1)$. Suppose $g = f_{r,k+3}$; then the function r_k which we require is given by

$$r_k(n,m) = t_{2,k+1}(r,n,m). \quad \square$$

Similarly (cf. Exercise 4.7) we may code the process of composing recursive functions (as in Theorem 4.1(c)) by a recursive function. We thus obtain

Proposition 4.9. There is a recursively enumerable set E such that
   (a) for every $n \in E$, $f_{n,1}$ is primitive recursive, and
   (b) every primitive recursive function of 1 variable occurs as $f_{n,1}$ for some $n \in E$.

Note that we do not assert that the set of *all* codes for primitive recursive unary functions is recursively enumerable. Nevertheless, in conjunction with the argument of Example 4.6, Proposition 4.9 suffices to prove our earlier assertion that not every globally defined recursive function is primitive recursive.

Proof. By definition, a function is primitive recursive iff it appears as the last term in a finite sequence $f_1,f_2,\ldots,f_t$, in which each term is either a basic function, or obtained by composing earlier terms in the sequence, or obtained by performing primitive recursion on two earlier terms. (We can think of such a sequence as a 'program for proving that f_t is primitive recursive'.) Clearly, we may code such 'programs' by natural numbers. Now we can give explicit codes for the basic functions, since we gave explicit programs for computing them in the proof of Theorem 4.1; thus we may describe an algorithm which, given a code for a sequence $(f_1,\ldots,f_t)$ as above, successively computes codes for each f_i, using the functions r_k of Lemma 4.8 and $c_{k,l}$ of Exercise 4.7 as 'subroutines'. By Church's Thesis, this algorithm corresponds to a recursive function, whose image is the required set E. $\square$

As another example of the combined power of Theorems 4.3 and 4.7, we give

Proposition 4.10. Let h be a globally defined recursive function of 1 variable. Then there is a number n such that $f_{n,1} = f_{h(n),1}$.

Proof. Consider the binary recursive function

$$g(x, y) = u(h(u(x, 1, 3^x)), 1, 3^y).$$

By Theorem 4.7, we can find a globally defined recursive function of 1 variable $(f_{m,1}$, say$)$ such that $g(x, y) = f_{f_{m,1}(x),1}(y)$. Let $n = f_{m,1}(m)$. Then for all y we have

$$
\begin{aligned}
f_{n,1}(y) &= f_{f_{m,1}(m),1}(y) \\
&= g(m, y) \\
&= u(h(f_{m,1}(m)), 1, 3^y) \\
&= u(h(n), 1, 3^y) \\
&= f_{h(n),1}(y),
\end{aligned}
$$

as required. $\square$

Proposition 4.10 is of most interest in the case when the function h is 'extensional' in the sense that, if n_1 and n_2 are codes for the same (unary) recursive function, then so are $h(n_1)$ and $h(n_2)$. We can then think of h as a 'recursive' mapping of the set of recursive functions into itself; and Proposition 4.10 tells us that every such mapping has a fixed point. (This is part of a general result called the Recursion Theorem, which implies that every such mapping has a *smallest* fixed point; i.e., among all the fixed points of the mapping, there is one with smallest possible domain, which is the restriction of all the others to that domain.) However, it should be noted that the fixed point $f_{n,1}$ may not be globally defined, even if h sends (codes for) globally defined functions to globally defined functions; for example, if h corresponds to the operation on programs which inserts a subroutine to add 1 to the final output, then the only fixed point of h is the empty function.

Next we fulfil a promise made in Chapter 1, by giving an example of an algebraic theory whose word problem is recursively insoluble. To do this, we must first 'code' the problem as a problem about natural numbers: for this, let us suppose that our operational type Ω is a countable set $\{\omega_0, \omega_1, \omega_2, \ldots\}$ and that the function which sends n to the arity $\alpha(\omega_n)$ of ω_n is recursive. Next, we suppose our set of variables is $X = \{x_0, x_1, x_2, \ldots\}$, and we code finite strings of elements of $\Omega \cup X$ by numbers of the form $2^{i_0} 3^{i_1} \ldots p_r^{i_r}$, where $i_j = 2n$ if the $(j+1)$th member of the string is ω_n, and $i_j = 2n+1$ if the $(j+1)$th member is x_n. Remark 1.1 can now be interpreted as a proof that the set of codes for Ω-terms is recursive.

Let us write $\ulcorner t \urcorner$ for the number coding a term t. We now further suppose that our set E of primitive equations is such that the set

$$\{(\ulcorner s \urcorner, \ulcorner t \urcorner) \mid (s = t) \in E\}$$

is recursive. An algebraic theory (Ω, E) satisfying the assumptions we have made so far is said to be *recursively presented*; we say that such a theory as a *(recursively) soluble word problem* if the set

$$\{(\ulcorner s \urcorner, \ulcorner t \urcorner) \mid (s = t) \in \tilde{E}\}$$

of codes for derived equations is also recursive.

Example 4.11. We shall construct a recursively presented theory (Ω, E) whose word problem is insoluble. Our operational type Ω will have one binary operation and a countable infinity of constants; our equations will include the associative law for the binary operation, and we shall find it convenient to denote this operation simply by juxtaposition. It will also be convenient to label our constants (in some recursive fashion) in two families ρ_m^n $(m, n \in \mathbb{N})$ and σ_m^n $(m, n \in \mathbb{N})$.

Among the Ω-terms we shall have strings like

$$\sigma_{i_0}^n \rho_{i_1}^n \rho_{i_2}^n \ldots \rho_{i_t}^n$$

which begin with a σ followed by a finite string of ρs, and in which all the *superscript* indices have the same value n; we shall call these *special terms* of index n, and our remaining primitive equations will all be between pairs of such special terms having the same index. We regard a special term as above as corresponding to a 'configuration' of our register machine when it is running the program P_n, namely that in which the program is in state S_{i_0} and the registers R_j hold the numbers i_j $(1 \leqslant j \leqslant t)$. We say a special term of index n has *adequate length* if it contains enough ρs to specify the contents of every register which can be affected by P_n.

Our next group of primitive equations is determined, for each n, by the instructions in the program P_n: for example, if the instruction to be obeyed in state S_3 of P_n is $(2, -, 4, 7)$, we write down all equations of the forms

$$(\sigma_3^n \rho_i^n \rho_0^n = \sigma_7^n \rho_i^n \rho_0^n) \quad \text{and} \quad (\sigma_3^n \rho_i^n \rho_{j+1}^n = \sigma_4^n \rho_i^n \rho_j^n).$$

Clearly, the set of (codes for) such equations is recursive. Let us now say that two configurations C_1 and C_2 are *n-equivalent* if the equality of the corresponding special terms of index n is derivable from the primitive equations we have so far; then it is clear that C_1 and C_2 are n-equivalent if the program P_n, starting from configuration C_1 (with some suitable values in the registers whose contents are not specified by C_1), eventually reaches C_2. The converse is false, because of the symmetric nature of equality: essentially, n-equivalence corresponds to what would happen if the machine were allowed to run backwards as well as forwards. However, because the forwards behaviour of the machine is deterministic (i.e. there is only one configuration to which it can move in a single step from a given configuration, although there may be several from which it can move to a given one), it is easy to see that two configurations C_1 and C_2 of adequate length are n-equivalent iff there is some third configuration C_3 which is eventually reachable by P_n from either of them.

Finally, we add to our theory all equations of the form

$$(\sigma_0^n \rho_{i_1}^n \rho_{i_2}^n \ldots \rho_{i_t}^n = \sigma_0^n);$$

this has no effect on the notion of n-equivalence for configurations from which P_n can never terminate, but has the effect of making all eventually-terminating configurations n-equivalent. Thus the program P_n, starting with input $(n, 0, 0, \ldots)$, will eventually terminate iff

$$(\sigma_1^n \rho_n^n \rho_0^n \rho_0^n \ldots \rho_0^n = \sigma_0^n)$$

is a derived equation of the theory we have constructed (the number of ρ_0^ns on the left-hand side being one less than the number of the remotest register affected by P_n – which is of course recursively computable from n). So a recursive solution of the word problem for this theory would lead to a recursive solution of the halting problem, contradicting Example 4.5.

To conclude this chapter, we shall investigate the relationship between Peano arithmetic, as defined in the last chapter, and recursive functions. To do this, we need to introduce a general concept of definability. Let T be a first-order theory (in a language $\mathscr{L}$, say) and M a T-model; then a (partial) function $f: M^k \to M$ is said to be *definable* in T (or T-definable) if there is a formula p of $\mathscr{L}$ with $(k + 1)$ free variables $(x_1, \ldots, x_k, y,$ say) which is provably functional (i.e. the sentence

$$(\forall x_1, \ldots, x_k, y, y')((p \wedge p[y'/y]) \Rightarrow (y = y'))$$

is provable in T) and whose interpretation in M is the graph of f (i.e. $(m_1, \ldots, m_k, n) \in [p]_M$ iff $(f(m_1, \ldots, m_k)$ is defined and) $f(m_1, \ldots, m_k) = n$). [Note: some authors would call this 'strong definability'.] If f is definable in T, we shall often denote a formula defining it by '$f(x_1, \ldots, x_k) = y$' (and indeed we shall treat '$f(x_1, \ldots, x_k)$' as if it were a term of $\mathscr{L}$). Our aim is to prove that every recursive function $\mathbb{N}^k \to \mathbb{N}$ is definable in PA.

In general, it is easy to see that the interpretation in M of any term of $\mathscr{L}$ is a T-definable function; in the particular case which concerns us, this means that every polynomial function $\mathbb{N}^k \to \mathbb{N}$ with natural number coefficients is PA-definable. But we may also define polynomials with negative or non-integral coefficients, provided they take natural numbers as values: for example, the function $n \mapsto \frac{1}{2}n(n + 1)$ is definable by the formula $(ayy = mxsx)$. Also, the division algorithm

$$(\forall x, y)(\neg(y = 0) \Rightarrow (\exists! z, t)((t < y) \wedge (x = amyzt)))$$

is provable in PA (by induction on x, with y as a parameter; here $(t < y)$ is an abbreviation for $(\exists u)(atsu = y)$); thus the function

$$r(m, n) = \text{remainder when } m \text{ is divided by } n \quad \text{if } n > 0,$$
$$\text{undefined} \qquad\qquad\qquad\qquad \text{if } n = 0$$

is PA-definable.

Let $v: \mathbb{N}^3 \to \mathbb{N}$ be the function $(m, n, p) \mapsto r(m, n(p + 1) + 1)$. By the above remarks, v is clearly PA-definable. We shall require the following lemma from elementary number theory.

Lemma 4.12. Given any finite sequence $(n_0, n_1, \ldots, n_k)$ of natural numbers, there exist numbers m_0 and m_1 such that $v(m_0, m_1, i) = n_i$ for all $i \leqslant k$.

Proof. First we choose m_1 to be a multiple of $k!$ which is greater than n_i for all $i \leqslant k$. Then the numbers $m_1(i + 1) + 1, 0 \leqslant i \leqslant k$, are pairwise coprime, for if d divides both $m_1(i + 1) + 1$ and $m_1(j + 1) + 1$ $(i \neq j)$, then it also divides

$$(j + 1)(m_1(i + 1) + 1) - (i + 1)(m_1(j + 1) + 1) = j - i,$$

and so (since $0 < |j - i| \leqslant k$) d divides m_1, whence $d = 1$. So, by the Chinese Remainder Theorem, the congruences

$$x \equiv n_i \bmod (m_1(i + 1) + 1) \quad (0 \leqslant i \leqslant k)$$

are simultaneously soluble; we let m_0 be any positive solution of these congruences. □

The above proof can be formalized within PA; this should be intuitively clear by now, although actually writing down such a formal proof would be fairly tedious. We can now prove

Theorem 4.13. Every recursive function is PA-definable.

Proof. We shall show that the class of PA-definable functions has the closure properties given in the statement of Theorem 4.1. (a) and (b) are obvious, since the basic functions are the interpretations of terms in the language of PA (indeed, of weak PA); and it is also clear that a composite of PA-definable functions is PA-definable. Now suppose that g is obtained from f by minimalization (as in Theorem 4.1(e)), and that f is definable by a formula $p(x_1, \ldots, x_k, x_{k+1}, y)$; then g is definable by

$$(p(x_1, \ldots, x_k, y, 0) \wedge (\forall y')((y' < y) \Rightarrow$$

$$(\exists z)(\neg(z = 0) \wedge p(x_1, \ldots, x_k, y', z)))).$$

Finally, suppose h is obtained from f and g by primitive recursion (as in Theorem 4.1(d)) and f and g are definable by formulae p and q. Consider the formula

$$(\exists z_1, z_2)(p(x_1, \ldots, x_k, v(z_1, z_2, 0))$$

$$\wedge (\forall t)((t < x_{k+1}) \Rightarrow q(x_1, \ldots, x_k, t, v(z_1, z_2, t), v(z_1, z_2, st)))$$

$$\wedge (y = v(z_1, z_2, x_{k+1}))).$$

Using Lemma 4.12, we see that this formula is satisfied by a $(k + 2)$-tuple of natural numbers $(n_1, \ldots, n_k, n_{k+1}, m)$ iff $h(n_1, \ldots, n_k, n_{k+1}) = m$. Thus h is PA-definable. □

Theorem 4.13 explains why, in defining the theory PA, we introduced addition and multiplication as primitive operations, as well as zero and successor. It is certainly not true that every recursive function is definable in weak PA: indeed, it can be shown that the addition map $\mathbb{N}^2 \to \mathbb{N}$ is not so definable. However, including addition and multiplication as primitives gives us 'just enough' to ensure the definability of the function v, which (as we have seen) is the key to proving the 'primitive recursion case' of Theorem 4.13.

It should also be mentioned that the converse of Theorem 4.13 is false: there exist PA-definable functions which are not recursive. For if E is a recursively enumerable set which is not recursive, then the function ϕ_E of Lemma 4.4(c) is recursive and hence PA-definable, say by $p(x, y)$; and then the formula

$$((y = 0) \wedge \neg p(x, y))$$

defines the function $\phi_{\mathbb{N}-E}$, which by hypothesis is not recursive.

Exercises

4.1. Describe the behaviour of the following programs:

(a)

(b)

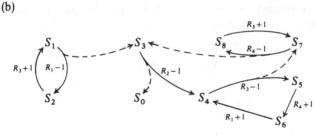

(c) $(1, -, 2, 4)$, $(3, +, 3)$, $(4, +, 1)$, $(2, -, 5, 7)$, $(5, +, 6)$, $(6, +, 4)$, $(3, -, 8, 9)$, $(5, -, 7, 16)$, $(5, -, 10, 18)$, $(5, +, 11)$, $(4, -, 12, 14)$, $(7, +, 13)$, $(3, +, 11)$, $(7, -, 15, 7)$, $(4, +, 14)$, $(6, -, 17, 0)$, $(2, +, 16)$, $(4, -, 19, 20)$, $(1, +, 18)$, $(1, +, 16)$.

4.2. (a) Write a program which computes $n^2 - 5n + 2$ if this is non-negative, and fails to terminate otherwise.

(b) Write a program which, given input n, computes the $(n + 1)$th prime number p_n.

4.3. Prove successively that the following functions are primitive recursive:

(i)   The function f_1 which is 1 if $n = 0$, and 0 otherwise.

(ii) $f_2(n) = $ the remainder when n is divided by 2.

(iii) $f_3(n) = $ the integer part of $n/2$.

(iv) $f_4(n) = n/2$ if n is even, 0 otherwise.

(v) $f_5(n, m) = n/2^m$ if this is an integer, 0 otherwise.

(vi) $f_6(n) = $ the exponent of the largest power of 2 dividing n, if $n > 0$,
    $ = 0$ if $n = 0$.

4.4. If E is a nonempty recursively enumerable subset of $\mathbb{N}$, show that there is a globally defined recursive function with E as its set of values.

4.5. If E is an infinite subset of $\mathbb{N}$, show that E is recursive iff there is a strictly increasing, globally defined recursive function $\mathbb{N} \to \mathbb{N}$ with E as its set of values.

4.6. If $f : \mathbb{N} \to \mathbb{N}$ is a (globally defined) recursive bijection, show that f^{-1} is also recursive.

4.7. Let k and l be positive integers. Show that there is a recursive function $c_{k,l}$ of $k + 1$ variables such that

$$f_{c_{k,l}(n,m_1,\ldots,m_k),l}(x_1, \ldots, x_l) = f_{n,k}(f_{m_1,l}(x_1, \ldots, x_l), \ldots, f_{m_k,l}(x_1, \ldots, x_l)).$$

4.8. Let S be a subset of the set of all unary recursive functions, such that both S and its complement are nonempty. Show that the set

$$\{n \in \mathbb{N} \mid f_{n,1} \in S\}$$

is not recursive. [Hint: let g be a function belonging to whichever of S and its complement does *not* contain the empty function, and consider the function h defined by

$$h(m, n) = g(n) \qquad \text{if } f_{m,1}(m) \text{ is defined,}$$
$$\text{undefined otherwise.}]$$

5

Zermelo–Fraenkel set theory

We now begin our formal development of set theory. We shall formulate it as a first-order theory in a language $\mathscr{L}$ whose only primitive symbol, apart from equality, is a binary predicate $\in$. (The axioms will, however, allow us to enrich $\mathscr{L}$ by introducing other primitive symbols later on.) A model of set theory will generally be denoted by the letter V; we think of V as being the 'universe' of all sets that we wish to talk about. Of course, V itself will be a set (or rather, will have an underlying set) in some 'meta-universe' which is clearly not identical with V; if we start trying to specify precisely what this meta-universe is we are clearly going to be led into problems of infinite regression, so we deliberately assume as little as possible about it. (Roughly, what we assume about the meta-universe is that it behaves in a way consistent with what we did in the first three chapters.)

We shall call the elements of V *sets*; and if $(a, b) \in [\in]_V \subseteq V^2$, we shall say that a is an *element* of b (in the model V). There is a possibility of confusion here, in that the words 'set' and 'element' now have technical meanings, when applied to elements of V, which may conflict with their meanings in the meta-universe; if we were very fussy we should probably introduce some formal means of distinguishing between the two uses of these words, but in fact it's usually possible to avoid confusion without going to such lengths, given common-sense and a modicum of goodwill. Note that, within our model, the elements of a set are themselves sets; we do not have a separate collection of 'ur-elements' which can occur on the left of the predicate $\in$ but not on the right (though there are some variants of set theory in which such things are allowed). Informally, the idea is that the entire universe can be built up inductively from the empty set $\emptyset$ by iterating the operation of forming collections of things. (We shall make this idea precise later on.)

Our fundamental picture of a set is that it is nothing more than the collection of its elements. This leads us to formulate our first axiom, the *axiom of Extensionality*

(Ext) $(\forall x, y)((\forall z)((z \in x) \Leftrightarrow (z \in y)) \Rightarrow (x = y))$

which says that two sets with the same elements are equal. (Note that the converse implication, that equal sets have the same elements, follows from our logical axioms for equality.)

Next, we want to be able to say that there exists a set whose elements are precisely the members of some well-defined collection. One appealing way to formalize this idea is the axiom-scheme of comprehension, which was (in essence) proposed by G. Frege in 1893:

$$(\exists y)(\forall x)((x \in y) \Leftrightarrow p)$$

for every formula p with $FV(p) = \{x\}$. Unfortunately, as Bertrand Russell pointed out, this scheme is inconsistent: by taking p to be the formula $\neg(x \in x)$, we deduce $(\exists y)((y \in y) \Leftrightarrow \neg(y \in y))$. Clearly, we have to be more careful about the formulae p that we allow. Russell's own solution to the problem was to reject set theory in favour of *type theory*, in which every entity we consider has a 'type' attached to it and the membership relation may only hold between entities of different types; thus the formula $\neg(x \in x)$ becomes impossible to interpret. However, in type theory we have to give up (or at least modify) the axiom of Extensionality, since there will exist empty sets of many different types. A variant of Russell's idea, which is associated with the name of W. Quine, is to retain the notion that we have just one type of entity (sets) but to restrict the comprehension-scheme to cases when the formula p is *stratified*; that is, when we can assign integer 'levels' to all the variables of p (both bound and free) in such a way that in any subformula $(x = y)$ the variables x and y have the same level, and in any subformula $(z \in t)$ the level of t is one greater than that of z.

However, in these notes we shall follow a different course, which was first proposed by E. Zermelo in 1908 (and formalized by T. Skolem in 1922), and which is now accepted by the great majority of set-theorists: namely, to restrict the use of the comprehension process to the construction of *subsets* of previously-constructed sets. In addition, we'll allow parameters in our formula p, as we did in the first-order induction-scheme for Peano arithmetic; thus we arrive at the following scheme, called the *axiom-scheme of Separation*:

(Sep) $(\forall z_1, \ldots, z_n)(\forall y_1)(\exists y_2)(\forall x)((x \in y_2) \Leftrightarrow ((x \in y_1) \wedge p))$

where p is any formula with $FV(p) = \{x, z_1, \ldots, z_n\}$. Note that (Ext) assures us that the value of y_2 making this sentence true (in a given model of set theory) is uniquely determined by the values assigned to $z_1, \ldots, z_n$ and y_1; so we may if we wish enrich our language by adding an $(n+1)$-ary operation-symbol S_p for each formula p as above, where

$$((y_2 = S_p(z_1, \ldots, z_n, y_1)) \Leftrightarrow (\forall x)((x \in y_2) \Leftrightarrow ((x \in y_1) \wedge p))).$$

Of course, we usually use the notation

$$\{x \in y_1 \mid p\}$$

for $S_p(z_1, \ldots, z_n, y_1)$; note that x appears in this notation as a bound variable.

Unfortunately, (Sep) on its own does not provide us with enough power to 'get started' in the business of constructing sets. The next four axioms are all instances of the comprehension-scheme (in fact stratified ones in the sense of Quine), which assert the existence of certain 'fundamental' sets, or of operations that we can perform on sets. Of course, now that we have (Sep), it is sufficient merely to assert the existence of some set containing the one we're interested in as a subset; thus each of these axioms can be presented in a 'strong' and a 'weak' form.

The first and simplest of them asserts that the empty set exists:

(Emp)$_s$ $(\exists x)(\forall y)\neg(y \in x)$.

The weak form (Emp)$_w$ of this axiom simply asserts that *some* set exists: $(\exists x)(x = x)$. In the presence of (Ext), we know that the empty set is unique, so we may add a constant $\varnothing$ to our language with the axiom $(\forall y)\neg(y \in \varnothing)$.

Next, we want an axiom to say that, given a set b, we can form the set whose only member is b. Actually, it is convenient for technical reasons to have an axiom about (unordered) pair-sets, from which the existence of singleton-sets follows as an easy consequence. The weak form of the *Pair-set axiom* says

(Pair)$_w$ $\qquad\qquad$ $(\forall x, y)(\exists z)((x \in z) \wedge (y \in z))$;

the strong form (Pair)$_s$ is

$$(\forall x, y)(\exists z)(\forall t)((t \in z) \Leftrightarrow ((t = x) \vee (t = y))).$$

The binary operation on V whose existence is implied by this axiom will of course be denoted by $\{x, y\}$; and we'll write $\{x\}$ as an abbreviation for $\{x, x\}$.

Thirdly in this group, we have the *Union axiom*

(Un)$_s$ $\qquad$ $(\forall x)(\exists y)(\forall t)((t \in y) \Leftrightarrow (\exists z)((t \in z) \wedge (z \in x)))$

(we leave it as an exercise to formulate the weak version), which says that the union of all the members of a set is a set. We write $\bigcup$ for the unary operation which this axiom allows us to introduce, and $(x \cup y)$ for $\bigcup \{x, y\}$. We shall similarly write $\bigcap x$ for the intersection of all the elements of x (provided $x \neq \varnothing$ – if x is empty then $\bigcap x$ doesn't exist!), but we don't need an extra axiom to introduce this one, since its existence follows from (Sep). (Similarly, (Sep) allows us to define the pairwise intersection $x \cap y$ and the set-theoretic difference $x - y$ of two sets x and y.)

The last axiom in this group is the *Power-set axiom*; we introduce the notation $(x \subseteq y)$ as a shorthand for $(\forall z)((z \in x) \Rightarrow (z \in y))$.

(Pow)$_s$ $\qquad\qquad$ $(\forall x)(\exists y)(\forall z)((z \in y) \Leftrightarrow (z \subseteq x))$.

We denote the unary operation introduced via this axiom by $\mathscr{P}$.

We now have enough axioms to do quite a lot of mathematics within our model V: for example, we can define the *ordered-pair* operation by

$$\langle x, y \rangle = \{\{x\}, \{x, y\}\}$$

(this trick is due independently to K. Kuratowski and N. Wiener), and prove that it has the property we expect of it:

$$(\forall x, y, z, t)((\langle x, y \rangle = \langle z, t \rangle) \Leftrightarrow ((x = z) \wedge (y = t))).$$

We can also define two unary operations

$$\text{First}(x) = \bigcup \bigcap x \qquad \text{if } x \neq \varnothing$$
$$= \varnothing \qquad\qquad \text{otherwise,}$$
$$\text{Second}(x) = \bigcup (\bigcup x - \bigcap x) \quad \text{if } (\bigcup x - \bigcap x) \neq \varnothing$$
$$= \text{First}(x) \qquad\quad \text{otherwise;}$$

then $\mathrm{First}(\langle x, y\rangle) = x$, $\mathrm{Second}(\langle x, y\rangle) = y$, and we can define the predicate 'is an ordered pair' by

$$(x \text{ is an ordered pair}) \Leftrightarrow (x = \langle \mathrm{First}(x), \mathrm{Second}(x)\rangle).$$

So we can define the cartesian product:

$$x \times y = \{t \in \mathscr{P}\mathscr{P} \bigcup \{x, y\} \mid (t \text{ is an ordered pair})$$
$$\wedge\, (\mathrm{First}(t) \in x) \wedge (\mathrm{Second}(t) \in y)\};$$

the predicate 'is a function':

$$(x \text{ is a function}) \Leftrightarrow ((\forall t)((t \in x) \Rightarrow (t \text{ is an ordered pair}))$$
$$\wedge\, (\forall y, z_1, z_2)(((\langle y, z_1\rangle \in x) \wedge (\langle y, z_2\rangle \in x))$$
$$\Rightarrow (z_1 = z_2)));$$

the ternary predicate 'is a function from ... to ...' (for which we use the familiar arrow notation):

$$(x : y \to z) \Leftrightarrow ((x \text{ is a function}) \wedge (\forall t)((t \in x) \Rightarrow (\mathrm{First}(t) \in y))$$
$$\wedge\, (\forall u)((u \in y) \Rightarrow (\exists v)((v \in z) \wedge (\langle u, v\rangle \in x))));$$

and the exponential:

$$z^y = \{x \in \mathscr{P}(y \times z) \mid x : y \to z\}.$$

Note, particularly in connection with the last but one of these, that it is convenient to introduce the *restricted quantifiers* $(\forall y \in x)p$ and $(\exists y \in x)p$ to mean, respectively,

$$(\forall y)((y \in x) \Rightarrow p) \quad \text{and} \quad (\exists y)((y \in x) \wedge p).$$

Observe that the variable x is free in these expressions, although y is bound; it is a condition of their use that x does not occur free in p.

Of course, when we say 'we can define' an n-ary operation symbol, we mean that there is an $(n + 1)$-ary predicate (p, say) in our original language (with only $=$ and $\in$), which is provably functional (i.e. the sentence

$$(\forall x_1, \ldots, x_n)(\exists! \, x_{n+1})p$$

is deducible from the axioms we have so far) and can be proved to have the properties we expect of it. Obviously, it would be unutterably tedious to reduce everything to formulae involving only $=$ and $\in$, which is why we feel free to enrich our language in this way, but the important thing is that it is *in principle* possible to do without the abbreviations we've introduced.

What we can't do from the first six axioms is to construct any infinite sets. We can certainly construct infinitely many distinct sets (for example, the sequence $\varnothing, \mathscr{P}\varnothing, \mathscr{P}\mathscr{P}\varnothing, \ldots$, or alternatively the sequence of (von Neumann) *natural numbers* defined inductively by

$$0 = \varnothing, \quad n+1 = n \cup \{n\},$$

so that the natural number n has exactly n elements), but we have no way of 'collecting' such sequences into infinite sets. 'Obviously', infinite collections form part of our mathematical intuition (though there are those who would argue against this), and so we need an axiom which allows us to do this. Such an axiom is the *axiom of Infinity*:

$(\text{Inf})_\text{w}$ $\qquad (\exists x)((\varnothing \in x) \wedge (\forall y)((y \in x) \Rightarrow (y^+ \in x)))$,

where we have introduced the notation y^+ for $y \cup \{y\}$ (the *successor* of y). A set x with the properties required by $(\text{Inf})_\text{w}$ is called a *successor set*; clearly, if a successor set exists then (Sep) allows us to find the unique smallest one, namely the intersection of all successor sets [exercise: formulate the strong axiom $(\text{Inf})_\text{s}$ yourself], and (Ext) tells us that this smallest successor set is unique, so that we can introduce a new constant ω to denote it.

Note that, in the presence of (Inf), the empty-set axiom (Emp) becomes redundant. Previously, it was the only axiom which unconditionally asserted the existence of a set, so that without it our universe V might have been empty. (Of course, if we eliminated (Emp) we'd have to eliminate the constant $\varnothing$ from the formula $(\text{Inf})_\text{w}$; but this is not hard to do.)

The seven axioms we have introduced so far, plus the axiom of Choice (which we shall discuss in Chapter 7 below), are essentially the axioms for set theory given by E. Zermelo in 1908. [*Note for historians*: Zermelo's axiom of infinity was slightly different from ours, since he defined the natural numbers by $0 = \varnothing, n+1 = \{n\}$; but this doesn't make any essential difference. Also, his separation axiom was not exactly like ours, because the concept of an arbitrary first-order formula didn't then exist and he referred to p as being a 'definite property' of sets, a phrase for which he never found a satisfactory definition.] However, subsequent workers in set theory realized the need for two more axioms, which have been added to the above seven to form what we now call *Zermelo–Fraenkel set theory* (ZF, for short).

The first of these extra axioms is the *axiom-scheme of Replacement*, which is traditionally associated with the name of A. Fraenkel, but which was in fact introduced independently by Fraenkel and by T. Skolem, about 1922. To see the need for this, let us return to the two infinite sequences $(\varnothing, \mathscr{P}\varnothing, \mathscr{P}\mathscr{P}\varnothing, \ldots)$ and $(0, 1, 2, \ldots)$ that we considered before introducing the axiom of Infinity. (Adopting part of a notation which will later become standard, we'll denote the former sequence by $(V_0, V_1, V_2, \ldots)$.) The axiom of Infinity tells us that we can 'collect' the terms of the second sequence into a single set ω; but it is not clear that we can do the same for the first, even though the sequences have 'the same number' of terms. More precisely, the 'function' $n \mapsto V_n$ is definable by a first-order formula (see Exercise 5.2) and its 'domain' is a set, but we don't yet know that its 'range' is a set.

Digression on sets and classes. We put the words 'function', 'domain' and 'range' into inverted commas in the last sentence, because $n \mapsto V_n$ is not (yet) a function *inside* our model V. In order to get over this difficulty, it's convenient to introduce the notion of a *class*; informally, a class is a sub-(meta-)set of V which is the interpretation of some first-order formula of our language. More formally, a class is an equivalence class of formulae with one free variable (x, say), the equivalence relation being that p and q define the same class if we can prove $(\forall x)(p \Leftrightarrow q)$ (in which case we say that p and q are *extensionally equivalent*). We'll use the informal notation $\{x \mid p\}$ for the class defined by a formula p; then $(y \in \{x \mid p\})$ is just another way of writing the formula $p[y/x]$. Also, we'll say the class $\{x \mid p\}$ *is a set* if we can prove

$$(\exists y)(\forall x)(p \Leftrightarrow (x \in y)); \tag{*}$$

if we can prove the negation of this formula, then we call $\{x \mid p\}$ a *proper class*. (Of course, we may not be able to prove either (*) or its negation, and indeed (*) may have different truth-values in different models of set theory.) In a similar way, we'll use the term *function-class* for (the extensional equivalence class of) a formula f with two free variables (x and y, say) for which we can prove

$$(\forall x, y, z)((f \wedge f[z/y]) \Rightarrow (y = z));$$

the *domain* of a function-class f is then the class $\{x \mid (\exists y)f\}$, and its *range* is the class $\{y \mid (\exists x)f\}$.

In terms of the concepts introduced in the last paragraph, we want an axiom which says that if the domain of a function-class is a set, then its range is a set. Of course, this involves a quantification over function-classes, and so we shall have to represent it as a scheme of first-order axioms, one for each formula defining a function-class. And once again it will be convenient to allow additional parameters in our formula; thus we arrive at the axiom-scheme

(Rep) $(\forall z_1, \ldots, z_n)((\forall x, y_1, y_2)((p \wedge p[y_2/y_1]) \Rightarrow (y_1 = y_2))$

$$\Rightarrow (\forall t)(\exists u)(\forall y_1)((y_1 \in u) \Leftrightarrow (\exists x)((x \in t) \wedge p))),$$

where p is any formula with $FV(p) = \{x, y_1, z_1, \ldots, z_n\}$. (If you find it hard to visualize what this is saying, think first about the case where there are no parameters, i.e. $n = 0$.) This formulation of the replacement-scheme is actually rather stronger than we need; in particular, it implies the separation-scheme (see Exercise 5.1). There are weaker versions of replacement which do not have this property, but they are (even) less intuitive than the one we've given.

Using Replacement, we can form the set which it is natural to denote informally by $\{V_n \mid n \in \omega\}$; and then we can form $\bigcup \{V_n \mid n \in \omega\}$, which we denote by V_ω. This is a fairly large set (albeit still countable); in fact it contains as a subset every particular set that we've mentioned so far. But the universe doesn't end here, since (by the Cantor diagonal argument) $\mathscr{P}V_\omega$ is strictly larger; it seems reasonable to denote $\mathscr{P}V_\omega$ by $V_{\omega+1}$, and to define $V_{\omega+2} = \mathscr{P}V_{\omega+1}, \ldots, V_{\omega+\omega} = \bigcup \{V_{\omega+n} \mid n \in \omega\}$, $V_{\omega+\omega+1} = \mathscr{P}V_{\omega+\omega}, \ldots$.

How long should we go on doing this? We'll defer the answer to this question until Chapter 6, and look instead at a related one: can we be sure that every set in our universe eventually turns up inside some V_α? [At the moment, 'α' is simply a generic name for the subscripts which appeared in the last paragraph; in the next chapter it will have a more specific meaning.] The answer (at present) is no, because we don't have anything to exclude the possibility that some set may be an element of itself, whereas it's easy to see that no such set can occur as a member of any V_α (consider the first V_α to which it belongs). To exclude this sort of pathological behaviour of the $\in$-predicate we need one further axiom, whose importance was first recognized (around 1925) by J. von Neumann, and which is called the *axiom of Foundation* (or *Regularity*).

The simplest formulation of this axiom is somewhat opaque: it says that every nonempty set has an element which is disjoint from itself, or in symbols

(Fdn) $(\forall x)(\neg(x = \varnothing) \Rightarrow (\exists y \in x)(y \cap x = \varnothing))$.

The idea is that, if we think of the universe as being built up in successive 'stages' represented by the V_αs, then among the members of a given nonempty set x there must be one which was constructed 'at least as early' as all the others; the members of this set must all have been constructed strictly earlier, and so cannot be members of x. In general, we shall say that a binary relation-class (defined by a formula $r(x, y)$ with two free variables) is *well-founded* if every nonempty set has an 'r-minimal' element (i.e.

$$(\forall z)(\neg(z = \varnothing) \Rightarrow (\exists y \in z)(\forall x)(r(x, y) \Rightarrow \neg(x \in z)))   );$$

so the axiom of Foundation is simply the assertion '$\in$ is well-founded'.

The chief use of the axiom of Foundation is to enable us to prove things about sets by induction over the membership relation $\in$. Before embarking on this topic, however, we need to introduce the concepts of transitive set and transitive closure. We say a set x is *transitive* if

$$(\forall y, z)(((z \in y) \wedge (y \in x)) \Rightarrow (z \in x));$$

it is easy to see that this is equivalent to either of the assertions $\bigcup x \subseteq x$ or $x \subseteq \mathscr{P}x$. (However, it is *not* the same as saying that the restriction of $\in$ to $x \times x$ is a transitive relation in the usual sense – the latter is equivalent to saying that every member of x is a transitive set.) It's not hard to see that an arbitrary intersection of transitive sets is transitive; hence if a given set x is contained in *some* transitive set, then there is a unique smallest transitive set which contains it, namely the intersection of all such sets. We call this smallest set the *transitive closure* of x, and denote it by TC(x); given (Un), (Inf) and (Rep), it may be constructed explicitly by

$$\mathrm{TC}(x) = \bigcup \{x, \bigcup x, \bigcup\bigcup x, \ldots\}.$$

We are finally ready to prove the first theorem of this chapter:

Theorem 5.1. In the presence of the other axioms of ZF, the axiom of Foundation is equivalent to the following axiom-scheme, known as the *principle of* $\in$-*induction*:

$$(\forall z_1, \ldots, z_n)((\forall x)((\forall y \in x)(p[y/x]) \Rightarrow p) \Rightarrow (\forall x)p), \quad (*)$$

where p is any formula with $\mathrm{FV}(p) = \{x, z_1, \ldots, z_n\}$.

Proof. First assume the axiom of Foundation. For simplicity, we shall consider the 'no-parameter' case of $(*)$, i.e. the case $n = 0$. Let p be some formula with $\mathrm{FV}(p) = \{x\}$, and assume the hypothesis that p is '$\in$-inductive', i.e.

$$(\forall x)((\forall y \in x)(p[y/x]) \Rightarrow p).$$

Assume further $\neg(\forall x)p$, i.e. $(\exists x)\neg p$. Given a particular x for which $\neg p$, consider the set

$$u = \{y \in \mathrm{TC}(\{x\}) \mid \neg p[y/x]\}.$$

u is nonempty since $x \in u$, and so it has an $\in$-minimal member y, say. But then we have $\neg p[y/x]$ since $y \in u$, and $(\forall z \in y)p[z/x]$ since the members of y are all in $\mathrm{TC}(\{x\}) - u$. This contradicts the $\in$-inductiveness of p; so from (p is $\in$-inductive) and $\neg(\forall x)p$ we have deduced $\bot$. Hence $((p$ is $\in$-inductive) $\Rightarrow (\forall x)p)$.

Conversely, suppose given the principle of $\in$-induction. We shall apply it to the predicate 'is a regular set', where 'x is a regular set' means

$$(\forall y)((x \in y) \Rightarrow (\exists z \in y)(z \cap y = \varnothing)).$$

Clearly, the axiom of Foundation is equivalent to saying that every set is regular, so it suffices to show that the property of being regular is $\in$-inductive. So suppose $(\forall y \in x)(y$ is regular), and let $x \in z$. Then either x itself is an $\in$-minimal member of z, or $x \cap z$ is nonempty. But in the latter case, any $y \in x \cap z$ is regular (being a member of x), and hence z has an $\in$-minimal member. $\quad \square$

Generalizing the first half of the proof of Theorem 5.1, we can prove a 'principle of R-induction' for any well-founded relation-class R, *provided* we have the analogue for R of the transitive closure operation. We can get this using the axioms of Infinity and Replacement, as before, provided we know that the 'R-predecessors' of any set form a set, i.e. provided

$$(\forall y)(\exists z)(\forall x)((x \in z) \Leftrightarrow (\langle x, y \rangle \in R)).$$

A relation-class with this property (which is of course trivial for $\in$) is said to be *local*. We can further modify the principle of R-induction by restricting the variables x and y to some class (or even a set), rather than the whole universe, provided R satisfies the well-foundedness condition relative to this class; note that if we restrict to a set, then the 'local' condition is automatic. If R is well-founded and local relative to a class M, we shall write RC_M for the R-closure operation relative to M; i.e. if x is a sub*set* of M, we write $RC_M(x)$ for the smallest set y satisfying $x \subseteq y$ and

$$(\forall z, t \in M)(((t \in y) \wedge (\langle z, t \rangle \in R)) \Rightarrow (z \in y)).$$

In particular, $\in C_V$ is the function-class which we have previously called TC.

Example 5.2. Let R be the successor-relation on ω (i.e.

$$(\langle x, y \rangle \in R) \Leftrightarrow ((x \in \omega) \wedge (y = x^+))\ ).$$

Then R is local (since each element of ω has at most one R-predecessor) and well-founded (since any $\in$-minimal member of a nonempty subset of ω is also R-minimal). Thus we obtain the usual *principle of mathematical induction*:

$$((p[\varnothing/x] \wedge (\forall x \in \omega)(p \Rightarrow p[x^+/x])) \Rightarrow (\forall x \in \omega)p)$$

where p is any formula with one free variable x (and the extension of this where p is allowed to contain parameters). Since every subset of ω is a subclass, this means that *within* our universe V ω is a model for the higher-order Peano postulates, as discussed in Chapter 3. Viewed from outside (i.e. as a set in our meta-universe), the class of natural numbers in V may not satisfy the higher-order induction postulate (since it may have sub-meta-sets which do not correspond to actual subsets of ω in V), but it is at least a model of first-order Peano arithmetic: the addition and multiplication maps $\omega \times \omega \to \omega$ may be defined by applying the argument of Exercise 3.8(b) internally in V. (We shall return to this example in Chapter 9.)

The natural next step after induction is recursion, which is concerned with defining things 'step by step' rather than proving them. Since we made extensive use of recursive definitions in the first four chapters, it is clearly desirable to prove a theorem saying that such definitions can be carried out within a model of ZF set theory. Before embarking on the proof of the R-Recursion Theorem, we need one technical preliminary: there is one point in the proof where we need to assume that our relation-class R is transitive (in the usual sense!), and so we require

Lemma 5.3. Let R be a relation-class which is well-founded and local relative to a class M. Then there exists a relation-class $\bar{R}$ containing R, which is also well-founded and local relative to M, and additionally transitive on M, i.e.

$$(\forall x, y, z \in M)(((\langle x, y \rangle \in \bar{R}) \land (\langle y, z \rangle \in \bar{R})) \Rightarrow (\langle x, z \rangle \in \bar{R})).$$

Proof. We define $\bar{R}$ by

$$(\langle x, y \rangle \in \bar{R}) \Leftrightarrow ((y \in M) \land (x \in RC_M(\{y\}))).$$

From the form of the definition, $\bar{R}$ is local; and it is transitive since $x \in RC_M(z)$ implies $RC_M(\{x\}) \subseteq RC_M(z)$. Suppose x is a nonempty subset of M with no $\bar{R}$-minimal member; then the set

$$\{y \in RC_M(x) \mid (\exists z \in x)(\langle z, y \rangle \in \bar{R})\}$$

is nonempty (since it contains x as a subset) and has no R-minimal member. So the well-foundedness of R implies that of $\bar{R}$. □

Theorem 5.4 (R-Recursion Theorem). Let M be a class, R a relation-class which is well-founded and local on M, and G a function-class (of two variables) which is defined everywhere on $M \times V$ (i.e. such that

$$(\forall x, y)((x \in M) \Rightarrow (\exists ! z)(z = G(x, y)))\quad).$$

Then there exists a unique function-class F, defined everywhere on M, satisfying

$$(\forall x \in M)(F(x) = G(x, \{F(y) \mid (y \in M) \land (\langle y, x \rangle \in R)\})). \quad (*)$$

The function-class F is said to be defined by *R-recursion* from G over the class M. A further generalization of the theorem allows parameters (i.e. additional variables, which may range over some class other than M) in F and G; their presence doesn't complicate the ideas of the proof, but only its notation.

Proof. Uniqueness is easy: if F_1 and F_2 both satisfy $(*)$, we may prove $(\forall x \in M)(F_1(x) = F_2(x))$ by R-induction over M. To prove existence, we introduce the notion of an *attempt* at defining F: by 'f is an attempt' we mean

(f is a function) $\wedge$ (dom$(f) \subseteq M$)

$\wedge\ (\forall x, y \in M)(((\langle x, y \rangle \in R) \wedge (y \in \text{dom}(f))) \Rightarrow (x \in \text{dom}(f)))$

$\wedge\ (\forall x \in \text{dom}(f))(f(x) = G(x, \{f(y) \mid (y \in M) \wedge (\langle y, x \rangle \in R)\}))$.

Our function-class F will be the 'union' of all possible attempts, i.e.

$$(F(x) = y) \Leftrightarrow (\exists f)((f \text{ is an attempt}) \wedge (f(x) = y)).$$

To prove that this is a function-class, we must show that any two attempts agree on the intersection of their domains; but this is just the uniqueness argument again, except that the induction is over this subset of M rather than the whole class.

To show that F is defined everywhere on M, we must prove

$$(\forall x \in M)(\exists f)((f \text{ is an attempt}) \wedge (x \in \text{dom}(f))).$$

Suppose this fails; let x be an element of M which is not in the domain of any attempt. Let x_0 be an $\bar{R}$-minimal member of

$$\{y \in RC_M(\{x\}) \mid \neg(y \in \text{dom}(F))\},$$

where $\bar{R}$ is defined as in the proof of Lemma 5.3; then all the members of $RC_M(\{x_0\})$, except x_0 itself, are in the domain of F, and

$$\{\langle u, v \rangle \mid ((u \in RC_M(\{x_0\}) \wedge (v = F(u)))\}$$

(which is a set by Replacement) is itself an attempt, f_0 say. We may now define

$$f_1 = f_0 \cup \{\langle x_0, G(x_0, \{f_0(y) \mid (y \in M) \wedge (\langle y, x_0 \rangle \in R)\})\rangle\},$$

and f_1 is an attempt with $x_0 \in \text{dom}(f_1)$, contradicting our hypothesis. So F is defined everywhere on M; and it clearly satisfies (∗). □

We shall meet numerous applications of the ∈-Recursion Theorem in the next chapter. For the present, we give an application of the r-Recursion Theorem, where r is a well-founded relation on a set. We need a new definition: we say a relation-class R is *extensional* if

$$(\forall x, y)((\forall z)((\langle z, x \rangle \in R) \Leftrightarrow (\langle z, y \rangle \in R)) \Rightarrow (x = y)).$$

Thus the axiom of Extensionality is just the assertion that ∈ is extensional.

Theorem 5.5 (Mostowski's Isomorphism Theorem). Let a be a set, and r a well-founded extensional relation on a. Then there exists a unique pair (b, f) where b is a transitive set and

$f:(a,r) \to (b, \in)$ is an isomorphism of sets-with-a-binary-relation.

Proof. We define f by r-recursion over a:

$$f(x) = \{f(y) \mid (y \in a) \wedge (\langle y, x \rangle \in r)\}$$

(i.e. we take $G(x, y) = y$ in the notation of Theorem 5.4), and we define $b = \{f(x) \mid x \in a\}$. Then it is clear from the definition that b is a transitive set, that f is surjective and that $(\langle x, y \rangle \in r)$ implies $(f(x) \in f(y))$. To establish the converse of the latter implication, it suffices to show that f is injective – which we have to do anyway. Consider the formula (with free variable x)

$$p(x): (\forall y \in a)((f(x) = f(y)) \Rightarrow (x = y));$$

we shall prove $(\forall x \in a)p(x)$ by r-induction. For if p holds for all the r-predecessors of x and $f(x) = f(y)$, then $\langle z, y \rangle \in r$ implies $f(z) \in f(x)$ and hence $f(z) = f(t)$ for some r-predecessor t of x, whence $z = t$, i.e. $\langle z, x \rangle \in r$. Similarly, $\langle z, x \rangle \in r$ implies $\langle z, y \rangle \in r$, so $x = y$ by extensionality of r.

To show the uniqueness of (b, f), suppose (b', f') is another such. By composing the inverse of f with f', we get an isomorphism $g:(b, \in) \to (b', \in)$; then an easy $\in$-induction shows $(\forall y \in b)(g(y) = y)$. So $b = b'$ and $f = f'$. $\square$

Exercises

5.1. Show that the Pair-set axiom and the axiom-scheme of Separation are deducible from the Empty-set and Power-set axioms and the axiom-scheme of Replacement. [Hint for the latter: to verify the instance of Separation corresponding to a formula p, consider the function-class which is the identity in so far as p holds, and undefined otherwise.]

5.2. Write down a first-order formula p, with one free variable f, to express the assertion that f is a function whose domain is a nonzero natural number, with $f(0) = \varnothing$ and $f(n^+) = \mathscr{P}f(n)$ for all n such that $n^+ \in \mathrm{dom}(f)$. Show that

$$(\forall n \in \omega)(\exists f)(p \wedge (\mathrm{dom}(f) = n^+)),$$

and deduce (without using Replacement) that the operation $n \mapsto V_n$ is representable by a function-class.

5.3. A class M is said to be *transitive* if

$$(\forall x, y)(((x \in y) \wedge (y \in M)) \Rightarrow (x \in M)).$$

If M is transitive and $(V, \in)$ is any model of set theory, show that the substructure $([M]_V, \in)$ of $(V, \in)$ satisfies the axiom of Extensionality, and that it satisfies each of the axioms of Empty set, Pair-set and Union iff $[M]_V$ is closed under the corresponding finitary operation on V. What further property of M do we need to get a similar result for the Power-set axiom?

5.4. The class HF of *hereditarily finite* sets is defined by

$$(x \in \mathrm{HF}) \Leftrightarrow (\forall y \in \mathrm{TC}(\{x\}))(y \text{ is finite}),$$

where 'y is finite' means that there exists a bijection from y to some natural number. Show that the class of hereditarily finite sets, when regarded (as in Exercise 5.3) as a substructure of V, satisfies all the axioms of ZF except the axiom of Infinity.

5.5. Which axioms of ZF are satisfied (a) by the class of *hereditarily countable* sets, and (b) by the class of *hereditarily small* sets, where 'y is countable' means that there exists an injection $y \to \omega$, and 'y is small' means that there is an injection from y to some set in the sequence $\omega, \mathscr{P}\omega, \mathscr{P}\mathscr{P}\omega, \ldots$? [Note: this use of 'small' is not standard terminology.]

5.6. Let $\sigma: V \to V$ be the permutation which interchanges $\varnothing$ and $\{\varnothing\}$ and leaves everything else fixed. Define a new binary relation $\in_1$ on V by

$$x \in_1 y \quad \text{iff} \quad \sigma(x) \in y.$$

Which of the axioms of ZF are satisfied in the structure $(V, \in_1)$? Would it make any difference if we used the relation '$x \in \sigma(y)$' instead of '$\sigma(x) \in y$'?

5.7. If x is a transitive set, show that $\bigcup x$ and $\mathscr{P}x$ are both transitive. Is either of the converse implications true?

5.8. Use the $\in$-Recursion Theorem to show that there is a unique function-class $\overline{\mathrm{TC}}$ such that

$$(\forall x)(\overline{\mathrm{TC}}(x) = x \cup \bigcup \{\overline{\mathrm{TC}}(y) \mid y \in x\}),$$

and show that $\overline{\mathrm{TC}}$ coincides with the transitive closure operation as defined in the text. Why is $\overline{\mathrm{TC}}$ unsatisfactory as a definition of the transitive closure operation?

5.9. If a is any subset of ω, show that $\in$ is well-founded and extensional as a binary relation on a. What does the Mostowski Isomorphism Theorem yield when applied to $(a, \in)$?

6

Ordinals and well-orderings

In this chapter we investigate a particular class of well-founded relations, namely those which are linear orderings. We begin with a few definitions.

Let $<$ be a binary relation on a set a. We say

$\quad <$ is *irreflexive* if $(\forall x \in a) \neg (x < x)$;
$\quad <$ is *antisymmetric* if $(\forall x, y \in a)((x < y) \Rightarrow \neg (y < x))$;
$\quad <$ is *transitive* if $(\forall x, y, z \in a)(((x < y) \wedge (y < z)) \Rightarrow (x < z))$;
$\quad <$ is *trichotomous* if $(\forall x, y \in a)((x < y) \vee (y < x) \vee (x = y))$;

and $<$ is a *linear* (or *total*) *order* on a if it satisfies all the above conditions. Actually, the second condition is redundant, since it is implied by the first and third. Moreover, a well-founded relation is always irreflexive (since if $x < x$ then $\{x\}$ has no $<$-minimal member), and a well-founded trichotomous relation is transitive (since if we have $x < y$ and $y < z$ but not $x < z$, then $\{x, y, z\}$ has no minimal member). Thus a well-founded relation is a linear order iff it is trichotomous; we call such a relation a *well-ordering* of the set a. Equivalently, a well-ordering of a is a linear ordering $<$ of a such that every nonempty subset of a has a (necessarily unique) $<$-least member. (We say that x is the least member of a subset b, rather than simply minimal, if $(\forall y \in b)((x < y) \vee (x = y))$.)

A linear ordering is clearly extensional, so we have an immediate special case of Mostowski's Theorem (5.5): if $(a, <)$ is a well-ordered set, then there exists a unique pair (b, f) where b is a transitive set well-ordered by $\in$ and $f: (a, <) \to (b, \in)$ is an order-isomorphism. We define a (von Neumann) *ordinal* to be a transitive set well-ordered (equivalently, linearly ordered) by $\in$; we shall adopt the usual convention of using Greek letters to denote ordinals. Mostowski's Theorem tells us that the ordinals form a system of representatives for the isomorphism classes of well-ordered sets; we shall refer to the unique ordinal isomorphic to a given well-ordered set $(a, <)$ as the *order-type* of $(a, <)$.

Do we know any examples of ordinals? Well, $\emptyset$ is (rather trivially) an ordinal; and in fact all the (von Neumann) natural numbers, as defined in Chapter 5, are ordinals. More generally,

Lemma 6.1. If α is an ordinal, then so is α^+ $(= \alpha \cup \{\alpha\})$.

Proof. If $x \in y \in \alpha^+$, then either $y \in \alpha$ or $y = \alpha$, and in either case we deduce $x \in \alpha \subseteq \alpha^+$; so α^+ inherits transitivity from α. The proof that it inherits trichotomy is equally easy. □

The set ω of all natural numbers is easily seen to be an ordinal; this is a particular case of the result that the union of any set of ordinals is an ordinal. To prove this, we need a couple of lemmas.

Lemma 6.2. Any member of an ordinal is an ordinal.

Proof. Let α be an ordinal, $\beta \in \alpha$. Since α is transitive, we also have $\beta \subseteq \alpha$, and hence β is linearly ordered by $\in$. To show β is transitive, suppose $\delta \in \gamma \in \beta$; then both δ and β are members of α, so we have one of $\delta \in \beta$, $\delta = \beta$ or $\beta \in \delta$. But the latter two possibilities imply that $\{\beta, \gamma, \delta\}$ has no $\in$-minimal member. □

Lemma 6.3. If α and β are ordinals, then either $\alpha \subseteq \beta$ or $\beta \subseteq \alpha$.

Proof. Suppose $\alpha \nsubseteq \beta$; let γ be the $\in$-least member of $\alpha - \beta$. We shall show $\gamma = \alpha \cap \beta$. For if $\delta \in \alpha \cap \beta$, then $\gamma \in \delta$ or $\gamma = \delta$ would imply $\gamma \in \beta$, and so (since both γ and δ are members of α) we must have $\delta \in \gamma$; and conversely, if $\delta \in \gamma$, then $\delta \in \alpha$ but $\delta \notin \alpha - \beta$, so $\delta \in \alpha \cap \beta$. Hence by Extensionality $\gamma = \alpha \cap \beta$. Thus $\alpha \nsubseteq \beta$ implies $\alpha \cap \beta = \gamma \in \alpha$; and similarly $\beta \nsubseteq \alpha$ implies $\alpha \cap \beta \in \beta$. So if neither inclusion holds, we have $\alpha \cap \beta \in \alpha \cap \beta$, contradicting Foundation. □

Corollary 6.4. (a) For ordinals α and β, $\alpha \subseteq \beta$ is equivalent to ($\alpha \in \beta$ or $\alpha = \beta$).

(b) If α and β are ordinals, then one of $\alpha \in \beta$, $\alpha = \beta$ or $\beta \in \alpha$ holds.

(c) If a is a nonempty set of ordinals, then $\bigcap a$ (is an ordinal and) is the $\in$-least member of a.

Proof. (a) If $\alpha \subseteq \beta$, then either $\alpha = \beta$ or $\beta \nsubseteq \alpha$, whence $\alpha \in \beta$ by the proof of Lemma 6.3. Conversely, $\alpha \in \beta$ implies $\alpha \subseteq \beta$ since β is transitive.

(b) follows from (a) and the statement of Lemma 6.3.

(c) By Foundation, a has an $\in$-minimal member; by (b) any such member is actually $\in$-least, and by (a) it equals $\bigcap a$, since it is contained in every member of a and equal to one of them. $\square$

Proposition 6.5. The union of any set of ordinals is an ordinal.

Proof. A union of transitive sets is transitive. If a is a set of ordinals, then the members of $\bigcup a$ are ordinals by Lemma 6.2; and by Corollary 6.4(b) they are linearly ordered by $\in$. $\square$

Note that, by Lemma 6.2 and Corollary 6.4(b), the class ON of all ordinals satisfies all the properties of an ordinal except that of being a set. Our next goal is to prove that ON is characterized by the closure properties of Lemma 6.1 and Proposition 6.5: for this, we need to introduce the notions of limit and successor ordinals. Let α be an ordinal; then either α has an $\in$-*greatest* member or it doesn't. If it does (say β), then we have

$$(\forall \gamma)((\gamma \in \alpha) \Leftrightarrow ((\gamma \in \beta) \vee (\gamma = \beta))),$$

i.e. $\alpha = \beta^+$, and we say α is a *successor*. If not, then we have $\alpha = \bigcup \alpha$ (recall that we always have $\bigcup \alpha \subseteq \alpha$, since α is transitive), and we say α is a *limit*. [According to this definition 0 is a limit, though it's sometimes convenient to exclude 0 from the class of limit ordinals; we shall always make this exclusion explicit when we need it.]

Theorem 6.6. Let M be any class satisfying

$$(\forall x)((x \in M) \Rightarrow (x^+ \in M))$$

and

$$(\forall y)((y \subseteq M) \Rightarrow (\bigcup y \in M)).$$

Then every ordinal is in M.

Proof. We show $(\forall \alpha)((\alpha \in \text{ON}) \Rightarrow (\alpha \in M))$ by $\in$-induction. Suppose $\alpha \in \text{ON}$; then $(\forall \beta \in \alpha)(\beta \in \text{ON})$ by Lemma 6.2, so the inductive hypothesis yields $(\forall \beta \in \alpha)(\beta \in M)$. But by the remarks above, we have either $(\alpha = \bigcup \alpha)$ or $(\exists \beta \in \alpha)(\alpha = \beta^+)$; and in either case we deduce $\alpha \in M$. $\square$

In proving things by $\in$-induction, or defining them by $\in$-recursion, over the class ON, it is often convenient to subdivide the argument into cases, to deal separately with 0, with successors and with nonzero limits. We shall see numerous examples of this in the pages which follow. (Incidentally, when α and β are ordinals, we shall sometimes write $\alpha < \beta$ (resp. $\alpha \leqslant \beta$) instead of $\alpha \in \beta$ (resp. $\alpha \subseteq \beta$).)

Now that we have the concept of ordinal, we can give a precise definition of the sequence of sets V_α (sometimes called the *von Neumann hierarchy*) which we mentioned in the last chapter. Explicitly, we define a function-class $(\alpha \mapsto V_\alpha)$ from ON to V by $\in$-recursion:

if $\alpha = \beta^+$ is a successor, $V_\alpha = \mathscr{P} V_\beta$;

if α is a limit, $V_\alpha = \bigcup \{V_\beta \mid \beta \in \alpha\}$.

[Note that the second clause of the definition includes the assertion $V_0 = \varnothing$. If we wanted to, we could combine the two clauses into one saying $V_\alpha = \bigcup \{\mathscr{P} V_\beta \mid \beta \in \alpha\}$.] While we're about it, let us also define the *rank function* by $\in$-recursion on V:

$$\text{rank}(x) = \bigcup \{\text{rank}(y)^+ \mid y \in x\}.$$

An easy induction shows that the rank of any set is an ordinal (and that the rank of any ordinal is itself); the rank function and the von Neumann hierarchy are related by

Proposition 6.7. For any set x and any ordinal α, we have

$$((x \in V_\alpha) \Leftrightarrow (\text{rank}(x) < \alpha))$$

and

$$((x \subseteq V_\alpha) \Leftrightarrow (\text{rank}(x) \leqslant \alpha)).$$

Proof. Since $\beta \leqslant \alpha$ iff $\beta < \alpha^+$ and $x \subseteq V_\alpha$ iff $x \in V_{\alpha^+}$, the second assertion follows from the first. For the left-to-right implication of the first assertion, we prove

$$(\forall \alpha \in \text{ON})(\forall x)((x \in V_\alpha) \Rightarrow (\text{rank}(x) < \alpha))$$

by $\in$-induction in α. Suppose $x \in V_\alpha$; if α is a limit then by definition we have $x \in V_\beta$ for some $\beta < \alpha$, so that $\text{rank}(x) < \beta$ by the inductive

hypothesis. If $\alpha = \beta^+$ is a successor, then we have
$$(\forall y \in x)(y \in V_\beta),$$
whence by the inductive hypothesis we obtain
$$(\forall y \in x)(\text{rank}(y) < \beta),$$
so that $\text{rank}(x) \leqslant \beta$ by the definition of the rank function. The proof of the right-to-left implication is a similar $\in$-induction, but this time in the variable x; we leave the details as an exercise. $\quad\square$

The von Neumann hierarchy gives rise to an appealing (if somewhat unstable-looking) 'picture' of the set-theoretic universe, which perhaps explains why it is usually denoted by the letter V.

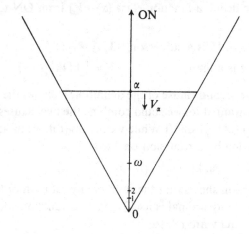

The idea of the picture is that the ordinals form a 'spine' running up the centre of the universe, and that the level at which any other set appears in the picture is determined by its rank. Thus Proposition 6.7 tells us that the set V_α is obtained by truncating the picture at level α. It is perhaps ironic that it should be the axiom of Foundation which tells us that the whole structure is balanced on a single 'point', namely the empty set. (However, the picture should not be taken too literally. It is certainly true in an informal sense that the universe 'gets wider as you go up', but the outer edges of the V do not exist in any meaningful sense.)

The von Neumann hierarchy can be defined even in set theory without the axiom of Foundation, since it follows from the definition of an ordinal that the restriction of $\in$ to ON is well-founded, even if $\in$ is not well-founded on V. In this context, the axiom of Foundation is equivalent to the assertion that every set lies in some V_α, or equivalently that the rank function is everywhere defined (cf. Exercise 6.2).

We now turn to the arithmetic of ordinals. There are two possible approaches to this subject, inductive and synthetic: for the former, we define the operations of addition and multiplication by $\in$-recursion over ON, and use $\in$-induction to prove that they satisfy certain laws (associativity, etc.), whereas in the latter we rely on Mostowski's Theorem, by defining $\alpha + \beta$ and $\alpha \cdot \beta$ as the order-types of certain well-ordered sets constructed from α and β. For completeness, we shall give (an outline of) both approaches.

The *ordinal sum* $\alpha + \beta$ is defined by $\in$-recursion on β (with α as a parameter):

$$\alpha + 0 = \alpha;$$

$$\alpha + \beta = (\alpha + \gamma)^+ \qquad \text{if } \beta = \gamma^+ \text{ is a successor;}$$

$$\alpha + \beta = \bigcup \{\alpha + \gamma \mid \gamma < \beta\} \quad \text{if } \beta \text{ is a nonzero limit.}$$

The *ordinal product* $\alpha \cdot \beta$ is also defined by recursion on β:

$$(\alpha \cdot 0 = 0);$$

$$\alpha \cdot \beta = \alpha \cdot \gamma + \alpha \qquad \text{if } \beta = \gamma^+ \text{ is a successor;}$$

$$\alpha \cdot \beta = \bigcup \{\alpha \cdot \gamma \mid \gamma < \beta\} \quad \text{if } \beta \text{ is a limit.}$$

(In this case the first clause of the definition isn't needed, because the third reduces to it when $\beta = 0$.)

Alternatively, we define $\alpha + \beta$ to be the order-type of the disjoint union $\alpha \amalg \beta$ (by which we mean, by definition, $\alpha \times \{0\} \cup \beta \times \{1\}$), ordered by

$$(\langle \gamma, i \rangle < \langle \delta, j \rangle) \Leftrightarrow ((i < j) \vee ((i = j) \wedge (\gamma < \delta)))$$

(i.e. the ordering in which all the elements of α precede all the elements of β; this is known as *reverse lexicographic ordering*, and it's easy to prove that r.l.o. on (a subset of) the product of two well-ordered sets is a well-ordering). $\alpha \cdot \beta$ is defined to be the order-type of the cartesian product $\alpha \times \beta$, ordered by r.l.o.

To verify that these two definitions agree, it is of course sufficient to show that the 'synthetic' definitions satisfy the various clauses of the recursive definitions. For example, if we take the successor case

$\beta = \gamma^+$ of multiplication, the synthetic definition tells us to form $\alpha \times \gamma^+ = (\alpha \times \gamma) \cup (\alpha \times \{\gamma\})$ with r.l.o.; but the elements of $\alpha \times \gamma$ precede those of $\alpha \times \{\gamma\}$ (and the latter have order-type α), so the inductive hypothesis that $\alpha \times \gamma$ has order-type $\alpha . \gamma$ implies that $\alpha \times \gamma^+$ has order-type $\alpha . \gamma + \alpha$. Similarly, if β is a limit, then we have $\alpha \times \beta = \bigcup \{\alpha \times \gamma \mid \gamma < \beta\}$, and each $\alpha \times \gamma$ forms an initial segment of $\alpha \times \beta$ in the r.l.o.; so if we have order-preserving bijections $\alpha \times \gamma \to \alpha . \gamma$ for each $\gamma < \beta$, we can put them together to form an order-preserving bijection $\alpha \times \beta \to \alpha . \beta$.

It follows immediately from either definition that $\alpha + 1 = \alpha^+$ for all α (so that we can now drop the notation α^+ for successors, if we wish). However, $1 + \alpha \neq \alpha + 1$ in general: for example, $1 + \omega = \bigcup \{1 + n \mid n \in \omega\} = \omega \neq \omega + 1$. Similarly, $2 . \omega = \omega$ but $\omega . 2 = \omega + \omega \neq \omega$. Thus ordinal addition and multiplication do not satisfy the commutative laws of ordinary (finite) arithmetic – which is not surprising when we consider the asymmetric nature of the definitions. The next question we must deal with, therefore, is which laws of arithmetic *do* hold for ordinals.

We begin with the order-preserving properties.

Lemma 6.8. (a) If $\beta < \gamma$, then $\alpha + \beta < \alpha + \gamma$.
  (b) If $\beta < \gamma$, then $\beta + \alpha \leqslant \gamma + \alpha$.
  (c) If $\alpha \geqslant 1$ and $\beta < \gamma$, then $\alpha . \beta < \alpha . \gamma$.
  (d) If $\beta < \gamma$, then $\beta . \alpha \leqslant \gamma . \alpha$.

Proof. Each of these may be proved either inductively or synthetically. In fact (a) and (c) are most easily proved synthetically, by showing that $\alpha \amalg \beta$ (resp. $\alpha \times \beta$) is a proper initial segment of $\alpha \amalg \gamma$ (resp. $\alpha \times \gamma$). (b) and (d) are most easily proved by induction on α: at a limit stage of the induction, we use the (obvious) fact that if each member of a set x is a subset of some member of y, then $\bigcup x \subseteq \bigcup y$. □

Next, some identities.

Lemma 6.9. The following hold for all ordinals α, β, γ:
  (a) $0 + \alpha = \alpha$,
  (b) $0 . \alpha = 0$,
  (c) $\alpha . 1 = \alpha = 1 . \alpha$,
  (d) $\alpha + (\beta + \gamma) = (\alpha + \beta) + \gamma$,

(e) $\alpha . (\beta + \gamma) = \alpha . \beta + \alpha . \gamma$,

(f) $\alpha . (\beta . \gamma) = (\alpha . \beta) . \gamma$.

Proof. Again, each of these can be proved by induction (except for the first half of (c), which follows at once from the definition and (a); the induction is on α in the first three cases, on γ in the other three), or synthetically by setting up an order-isomorphism between two well-ordered sets. We give one example of each method.

For the synthetic proof of (e), we must set up an order-isomorphism between $\alpha \times (\beta \amalg \gamma)$ and $(\alpha \times \beta) \amalg (\alpha \times \gamma)$. But the first of these sets has elements of the form $\langle \delta, \langle \eta, i \rangle \rangle$ where $\delta \in \alpha$ and either ($\eta \in \beta$ and $i = 0$) or ($\eta \in \gamma$ and $i = 1$); and the second has elements of the form $\langle \langle \delta, \eta \rangle, i \rangle$ subject to the same restrictions. It is clear that the unique sensible bijection between these two sets preserves r.l.o.

For the inductive proof of (f), suppose $\alpha . (\beta . \delta) = (\alpha . \beta) . \delta$ for all $\delta < \gamma$. If $\gamma = \delta^+$ is a successor, then we have

$$\alpha . (\beta . \gamma) = \alpha . (\beta . \delta + \beta) \quad \text{by definition}$$
$$= \alpha . (\beta . \delta) + \alpha . \beta \quad \text{by (e)}$$
$$= (\alpha . \beta) . \delta + \alpha . \beta \quad \text{by inductive hypothesis}$$
$$= (\alpha . \beta) . \gamma \quad \text{by definition.}$$

If γ is a limit, then $\alpha . (\beta . \gamma) = \alpha . \bigcup \{\beta . \delta \mid \delta < \gamma\}$; but the sequence of ordinals $\beta . \delta$ ($\delta < \gamma$) is strictly increasing by 5.8(c) (except in the case $\beta = 0$, which is trivial), from which it follows easily that its union must be a limit. Moreover, since any η less than this limit is less than some $\beta . \delta$, we actually have

$$\alpha . (\beta . \gamma) = \bigcup \{\alpha . \eta \mid \eta < \bigcup \{\beta . \delta \mid \delta < \gamma\}\}$$
$$= \bigcup \{\alpha . (\beta . \delta) \mid \delta < \gamma\}$$
$$= \bigcup \{(\alpha . \beta) . \delta \mid \delta < \gamma\} \quad \text{by inductive hypothesis}$$
$$= (\alpha . \beta) . \gamma \quad \text{by definition.} \quad \square$$

However, in addition to the two commutative laws, the 'right distributive law' $(\alpha + \beta) . \gamma = \alpha . \gamma + \beta . \gamma$ does not hold for ordinals: for example, take $\alpha = \beta = 1$, $\gamma = \omega$.

Exercises

6.1.   Some writers on set theory define an ordinal to be a transitive set
whose members are all transitive sets. By Lemma 6.2, the definition
given in the text implies this one; use $\in$-induction to prove that the
converse holds. Show also that the converse does *not* hold if we delete
Foundation from our axioms. [Consider the model of Exercise 5.6.]

6.2.   Show that, if we do not assume the axiom of Foundation, we still have

$$(\forall x)((x \text{ is regular}) \Leftrightarrow (\exists \alpha \in \text{ON})(x \in V_\alpha))$$

where 'x is regular' is defined as in the proof of Theorem 5.1.

6.3.   Show that $\text{rank}(x) = \text{rank}(\text{TC}(x))$ and that

$$\text{rank}(x) = \{\text{rank}(y) \mid y \in \text{TC}(x)\}$$

for all sets x.

6.4.   Show that any member of $V_{\omega+1} - V_\omega$ is infinite, and deduce that the
class HF defined in Exercise 5.4 is exactly the set V_ω. Show also that the
class HC of hereditarily countable sets (see Exercise 5.5) is a subset of
V_α for a suitably chosen α.

6.5.   Define a binary operation $\dot{-}$ on ordinals such that
(a) $\alpha \dot{-} \beta = 0$ if $\alpha < \beta$, and
(b) $\beta + (\alpha \dot{-} \beta) = \alpha$ if $\alpha \geqslant \beta$.
Give an example of a pair of ordinals (α, β) with $\alpha > \beta$, for which there
does not exist γ with $\gamma + \beta = \alpha$.

6.6.   (The division algorithm for ordinals.) Let α and β be ordinals, $\beta \neq 0$.
Show that there exists a unique pair (γ, δ) such that $\alpha = \beta \cdot \gamma + \delta$ and
$\delta < \beta$. [Hint: first show that there exists γ' such that $\alpha < \beta \cdot \gamma'$, and that
the least such γ' is a successor.]

6.7.   (a) Let α and β be ordinals. Show that the set F of all functions $\beta \to \alpha$ is
linearly ordered by lexicographic ordering (i.e.

$$(f < g) \Leftrightarrow (\exists \gamma \in \beta)((f(\gamma) < g(\gamma)) \wedge (\forall \delta < \gamma)(f(\delta) = g(\delta))) \quad),$$

but is not well-ordered unless β is finite.
(b) A function $f: \beta \to \alpha$ is said to have *finite support* if $\{\gamma \in \beta \mid f(\gamma) \neq 0\}$
is finite; let $F_0 \subseteq F$ be the subset of functions of finite support. Show
that F_0 is still not well-ordered by lexicographic ordering, but that it is
well-ordered by reverse lexicographic ordering (i.e. $f < g$ iff $f(\gamma) < g(\gamma)$
for the *largest* γ at which they differ).

6.8.   We define the *ordinal exponential* α^β, for ordinals α and β, by recursion in β, as follows:

$$\alpha^0 = 1$$
$$\alpha^\beta = \alpha^\gamma . \alpha \qquad \text{if } \beta = \gamma^+ \text{ is a successor}$$
$$\alpha^\beta = \bigcup \{\alpha^\gamma \mid \gamma < \beta\} \quad \text{if } \beta \text{ is a nonzero limit.}$$

(a) Prove that α^β is the order-type of the well-ordered set $(F_0, \text{r.l.o.})$ defined in Exercise 6.7(b).

(b) Prove that the laws $\alpha^{\beta+\gamma} = \alpha^\beta . \alpha^\gamma$ and $\alpha^{\beta \cdot \gamma} = (\alpha^\beta)^\gamma$ hold for all α, β, γ.

(c) Find a triple (α, β, γ) such that $(\alpha\beta)^\gamma \neq \alpha^\gamma . \beta^\gamma$.

6.9. A function-class $F : \mathrm{ON} \to \mathrm{ON}$ is called a *normal function* if

(a) $(\forall \alpha, \beta)((\alpha < \beta) \Rightarrow (F(\alpha) < F(\beta)))$, and

(b) $(\forall x \subseteq \mathrm{ON})((x \neq \varnothing) \Rightarrow (F(\bigcup x) = \bigcup \{F(\alpha) \mid \alpha \in x\}))$ (equivalently, F is 'continuous' at limits).

   If F is a normal function, show that $\alpha \leqslant F(\alpha)$ for all $\alpha \in \mathrm{ON}$, and that

$$\bigcup \{\alpha, F(\alpha), F(F(\alpha)), \ldots\}$$

is the least $\beta \in \mathrm{ON}$ satisfying $\alpha \leqslant \beta = F(\beta)$. Hence show that there exists a normal function $G : \mathrm{ON} \to \mathrm{ON}$ (called the *derivative* of F) such that the values of G are precisely the fixed points of F.

   If β and γ are ordinals with $\gamma \neq 0$, show that the functions F_β and G_γ defined by $F_\beta(\alpha) = \beta + \alpha$ and $G_\gamma(\alpha) = \gamma . \alpha$ are normal. What are their derivatives?

6.10. Show that the function $\alpha \mapsto \omega^\alpha$ (defined as in Exercise 6.8) is a normal function in the sense of Exercise 6.9. Hence show that every nonzero ordinal α has a unique representation of the form

$$\alpha = \omega^{\alpha_0} . a_0 + \omega^{\alpha_1} . a_1 + \ldots + \omega^{\alpha_n} . a_n,$$

where $\alpha \geqslant \alpha_0 > \alpha_1 > \ldots > \alpha_n$ and $a_0, a_1, \ldots, a_n$ are nonzero natural numbers. [This representation is called the *Cantor Normal Form* of α.]

7
The axiom of choice

We saw in the last chapter that ordinals (or equivalently well-orderings) are useful things to have around. Moreover, any model of set theory contains 'arbitrarily large' ordinals: not only is the class ON not a set (if it were, it would be a member of itself by Lemma 6.2 and Corollary 6.4(b), contradicting the well-foundedness of ordinals – this is known as the Burali–Forti paradox), but we have

Lemma 7.1 (Hartogs' Lemma). For any set a, there exists an ordinal α which cannot be mapped injectively into a.

Proof. α can be mapped injectively into a iff α is the order-type of some well-ordering of a subset of a. We can form the set $s \subseteq \mathscr{P}(a \times a)$ of all well-orderings of subsets of a using the Power-set and Separation axioms, and Mostowski's Theorem in fact produces a function-class which sends every well-ordering to its order-type; so by Replacement the ordinals which can be mapped injectively into a form a set $\gamma(a)$, say. $\gamma(a)$ is clearly an initial segment of ON, and hence is itself an ordinal; and it cannot be mapped injectively into a. $\square$

Nevertheless, for many purposes we would like something better than this: namely a bijection between a and an ordinal, or equivalently a well-ordering of the whole of a. The need for such became apparent in the last years of the nineteenth century, when G. Cantor began to prove results in point-set topology by transfinite induction methods. In 1904 E. Zermelo published his 'Beweis, dass jede Menge wohlgeordnet werden kann' (Proof that every set can be well-ordered), in which he used for the first time (explicitly, at any rate) a new principle about sets which became known as the axiom of choice.

This axiom immediately attracted controversy, essentially because it differs from the other axioms of set theory [apart from Foundation] in asserting the existence of something which is *not* specified exactly in terms of its members, but which can only be produced by making arbitrary choices. Explicitly, the axiom says that if $(a_i \mid i \in I)$ is a family of sets (indexed by a set I) and for each $i \in I$ a_i is nonempty, then there exists a function $f: I \to \bigcup \{a_i \mid i \in I\}$ with the property that $f(i) \in a_i$ for each $i \in I$. As a particular case, the axiom asserts that for any set a there is a function $g: (\mathscr{P}a - \{\varnothing\}) \to a$ such that $g(b) \in b$ for each $b \in \mathrm{dom}(g)$; but this particular case implies the general one by taking $a = \bigcup \{a_i \mid i \in I\}$. A function g as above is called a *choice function* for the set a.

Clearly, if I is finite we do not need any new axiom to define a function f as in the last paragraph; the case of interest is thus when I is infinite. Even here there may, in particular cases, be some 'uniformity' about the sets a_i which enables us to define f without making infinitely many choices: for example, if we had an infinite set of pairs of shoes, we could define the function which chooses the left shoe from each pair. (But if we had an infinite set of pairs of socks, things would be more difficult.)

Because of its controversial nature, mathematicians generally prefer not to regard the axiom of choice as one of the basic axioms of set theory, but to invoke it only when it is clear that the result they are aiming at cannot be achieved without it. Thus a great deal of work has been done on studying set-theoretical and mathematical statements which are equivalent (in the presence of the axioms of ZF set theory) to the axiom of choice, or to weak versions of it. One such is the result for which Zermelo introduced the axiom: the well-ordering theorem.

Proposition 7.2. For a set a, the following are equivalent:
   (i) a can be well-ordered.
   (ii) a has a choice function.

Proof. Given a well-ordering $<$ of a, we may define a choice function $g: (\mathscr{P}a - \{\varnothing\}) \to a$ by $g(b) = $ $<$-least element of b.
   Conversely, let $g: (\mathscr{P}a - \{\varnothing\}) \to a$ be a choice function. If $a = \varnothing$ we have nothing to prove; otherwise, we define a function-class

$F: \mathrm{ON} \rightarrow a$ by recursion:

If $\{F(\beta) \mid \beta < \alpha\} \neq a$, then $F(\alpha) = g(a - \{F(\beta) \mid \beta < \alpha\})$;
otherwise $F(\alpha) = g(a)$ ($= F(0)$).

By Lemma 7.1, there exists $\gamma \in \mathrm{ON}$ such that the restriction of F to γ isn't injective. But the only way it could fail to be injective is for the second clause of the definition to come into operation, i.e. there must exist $\alpha < \gamma$ with $\{F(\beta) \mid \beta < \alpha\} = a$. Consider the least such α: then $F|_\alpha$ is bijective, so we may transfer the well-ordering of α along it to get a well-ordering of a. $\square$

Remark. Proposition 7.2 is usually stated in the 'global' version

(AC) $\Leftrightarrow$ (Every set can be well-ordered);

but it's sometimes useful to have the 'local' version which applies to a particular set a, even when AC doesn't hold globally. Our proof of (ii) $\Rightarrow$ (i) is not the same as Zermelo's original proof, since we used Replacement (which wasn't available to Zermelo) in the proof of Lemma 7.1; Zermelo's method is sketched in Exercise 7.2 below.

In applying the axiom of choice to other areas of mathematics, the well-ordering theorem is often useful (and was extensively used by mathematicians in the 1920s and 1930s); but it has now been largely superseded by the result generally known as Zorn's Lemma or Hausdorff's Maximal Principle. The latter name is more correct, since Hausdorff published a version of it in 1909, some 26 years before Zorn; but it attracted little attention at the time, and it was not until Zorn's paper was published that the maximal principle began to be widely used as a substitute for the well-ordering theorem. (Also, Hausdorff has many other things named after him, whereas Zorn's only claim to fame is his Lemma.)

Before we state it, a bit of terminology: a *chain* in a partially ordered set $(a, \leqslant)$ is a subset b which is linearly ordered by the restriction of $<$ (i.e. $\leqslant$ and $\neq$) to $b \times b$. We say $(a, \leqslant)$ is *inductive* if every chain b in a has an upper bound (in a, not necessarily in b). Note that, since $\varnothing$ is a chain, an inductive p.o. set must in particular be nonempty. A *maximal* element of $(a, \leqslant)$ is an element m such that $(\forall x \in a)((m \leqslant x) \Rightarrow (m = x))$. Then Zorn's Lemma asserts

(ZL)   Every inductive partially ordered set has a maximal element.

Proposition 7.3. AC is equivalent to ZL.

Proof. Assume AC holds: let $(a, \leqslant)$ be an inductive p.o. set with no maximal element. Let c be the set of chains in a; by AC, there exist functions

$$g: a \to a \quad \text{such that } (\forall x \in a)(x < g(x))$$

and $\quad h: c \to a \quad$ such that $(\forall y \in c)(h(y)$ is an upper bound for $y)$.

Now define $F: \mathrm{ON} \to a$ recursively by

$$F(0) = h(\varnothing)$$
$$F(\alpha) = g(F(\beta)) \qquad\qquad \text{if } \alpha = \beta^+$$

and $\quad F(\alpha) = g(h(\{F(\beta) \mid \beta < \alpha\})) \quad$ if α is a nonzero limit ordinal.

An easy induction shows that F is strictly order-preserving, and in particular injective; but this contradicts Lemma 7.1.

Conversely, suppose ZL holds and let a be any set. Let b be the set of *partial choice functions* for a (i.e. functions f defined on a subset of $(\mathcal{P}a - \{\varnothing\})$ and satisfying $f(c) \in c$ for all $c \in \mathrm{dom}(f)$). We may partially order b by setting $f \leqslant g$ iff g extends f; then $(b, \leqslant)$ is inductive since the union of a chain of partial choice functions is a p.c.f. (and in particular the empty function is a p.c.f.). So b has a maximal element f_0, say. Suppose $\mathrm{dom}(f_0) \neq (\mathcal{P}a - \{\varnothing\})$; then by choosing $c \in (\mathcal{P}a - \{\varnothing\} - \mathrm{dom}(f_0))$ and $d \in c$ we may define $f_1 = f_0 \cup \{\langle c, d \rangle\}$. Now $f_1 \in b$ and $f_0 < f_1$, contradicting maximality of f_0; so f_0 must be a global choice function for a. $\square$

The style of the second half of the proof of Proposition 7.3 is entirely typical of arguments involving Zorn's Lemma. For example, one can give a direct proof of the well-ordering theorem from ZL, by applying it to the set of all partial well-orderings of a set a ($=$ well-orderings of subsets of a), partially ordered by the relation 'is an initial segment of'. We shall meet some further examples below.

Next, we list some examples of the applications of AC in areas of mathematics other than set theory. One of the most famous comes from general topology:

Theorem 7.4 (Tychonoff's Theorem). If AC holds, than an arbitrary cartesian product of compact topological spaces is compact.

This was proved by A. N. Tychonoff in 1929 using well-ordering; the proof involves too many topological ideas to be given here. In 1950 J. L. Kelley showed that the converse of Tychonoff's Theorem holds; his argument is quite simple, and is sketched in Exercise 7.3 below.

Next, one from ring theory:

Theorem 7.5 (Maximal Ideal Theorem). If AC holds, then any proper ideal in a ring (with 1) is contained in a maximal (proper) ideal.

Proof. Apply ZL to the set of proper ideals which contain the given one. To verify that this set is inductive, we need to show that the union of a chain of proper ideals is proper; this is true because an ideal is proper iff it doesn't contain 1. □

The Maximal Ideal Theorem was first proved (using well-ordering) by W. Krull in 1927; the fact that it too has a converse was proved much more recently (in 1979, in fact) by W. Hodges.

Theorem 7.6 (Hamel's Theorem). If AC holds, then any vector space has a basis.

Proof. Apply ZL to the set of linearly independent subsets of the space, ordered by inclusion. Since a set is l.i. iff each of its finite subsets is l.i., it's easy to show that this is inductive. The proof that a maximal l.i. set must be a spanning set is straightforward. □

This result was proved by W. Hamel in 1905 (for the particular case of $\mathbb{R}$ as a vector space over $\mathbb{Q}$); it is thus one of the earliest applications of AC outside set theory. Again, the converse is much more recent: in 1966 J. D. Halpern proved that the familiar assertion 'Any linearly independent set in a vector space can be extended to a basis' implies AC (part of his argument is sketched in Exercise 7.11 below), and in 1983 A. R. Blass proved that 'Every vector space has a basis' implies AC. Halpern's proof allows you to fix the underlying field before you start, but Blass's doesn't; thus it is still an open problem whether, for example, 'Every vector space over $\mathbb{R}$ has a basis' implies AC.

The last three results may all be considered 'pleasant' consequences of AC, in that they are important and useful tools in the development of the subjects to which they belong. (The same applies to the result, fundamental to the development of Galois theory, that AC implies that every field has an algebraic closure.) But in other areas the effect of AC can be less benign: a good example is measure theory, where AC implies the existence of subsets of $\mathbb{R}$ which are not Lebesgue measurable. (It was shown in 1970 by R. M. Solovay that if we are willing to jettison AC – and to make certain other set-theoretic assumptions – then we can find a model of ZF set theory in which every set of real numbers is measurable.)

A more spectacular 'unpleasant' consequence of AC is the so-called Banach–Tarski paradox. Banach and Tarski showed that, assuming AC, it is possible to decompose the unit ball in $\mathbb{R}^3$ into a finite number of disjoint pieces, and then reassemble two copies of the unit ball from congruent copies of these pieces (i.e. sets onto which the original pieces can be mapped by Euclidean motions of $\mathbb{R}^3$). [The argument is too complicated to give here; there is a good account of it on pp. 3–6 of *The Axiom of Choice* by T. J. Jech (North-Holland, 1973).]

Because of paradoxical results like this, mathematicians have devoted a good deal of study to weaker 'choice principles' which do not imply AC but which may be sufficient for the development of particular areas of mathematics. One of the most important is the *(Boolean) Prime Ideal Theorem*, which is the assertion that every nontrivial Boolean algebra has a homomorphism into 2. (A *prime ideal* in a Boolean algebra is a subset satisfying closure conditions which make it the kernel of such a homomorphism, i.e. the set of elements mapped to 0. The maximal consistent set S' which we constructed in the proof of the Completeness Theorem (2.5) is thus the complement of a prime ideal – such a set is called a *prime filter* – and in fact (see Exercise 7.5) the Completeness Theorem for Propositional Logic (*within* a given model of set theory) is equivalent to the Prime Ideal Theorem.)

D. S. Scott showed in 1954 that the Prime Ideal Theorem is equivalent to the assertion 'Every proper ideal in a commutative ring (with 1) is contained in a prime ideal' (cf. Theorem 7.5 above; recall that maximal ideals in a commutative ring are prime, but not in general conversely); and J. Łoś and C. Ryll-Nardzewski showed at about the same time that it is equivalent to 'An arbitrary cartesian product of compact Hausdorff spaces is compact' (cf. Theorem 7.4).

Both of these results have recently (1983–4) been extended, as a result of work by B. Banaschewski, A. R. Blass and P. T. Johnstone: they have shown that the word 'commutative' may be omitted from the ring-theoretic assertion, and that the word 'Hausdorff' in the topological one may be replaced by the much weaker condition 'sober'.

The Prime Ideal Theorem (PIT) implies that every set can be linearly ordered (cf. Exercise 7.6), which in turn implies the axiom of choice for families of finite sets (cf. Exercise 7.8). Neither of these implications is reversible, and the implication (AC ⇒ PIT) is also irreversible; the latter fact was first proved by J. D. Halpern in 1963. (We shall have a little to say about the technology of proving such 'independence' results in Chapter 9.)

Another obvious way of restricting the axiom of choice, apart from restricting the size of the sets in the families for which we demand choice functions, is to restrict the size of the families themselves. We have already remarked that we don't need any extra axiom to make a finite number of choices, so the first interesting case of this is the assertion that every countable family of nonempty sets has a choice function. Actually, this 'axiom of Countable Choice' is a little too weak for most of the situations where we need to make countably many arbitrary choices, since we often want to make them sequentially, in such a way that the range of values from which we make the nth choice depends on the preceding $n - 1$ choices. A way of expressing this is the *axiom of Dependent Choices*, formulated by P. Bernays in 1942:

(DC) If r is a binary relation on a set a such that

$$(\forall x \in a)(\exists y \in a)(\langle x, y \rangle \in r)$$

and b is any element of a, then there exists $f: \omega \to a$ such that $f(0) = b$ and $(\forall n \in \omega)(\langle f(n), f(n^+) \rangle \in r)$.

Uses of DC are quite common in mathematics, often to prove that two definitions of a concept are equivalent. A good example is given by the two definitions of a Noetherian ring, as a ring in which every ideal is finitely generated, or as a ring in which every increasing sequence of ideals is eventually constant: if we are given a strictly increasing sequence of ideals, then the union of the terms of the sequence is an ideal which cannot be finitely generated, but to go in the opposite direction from a non-finitely-generated ideal to an increasing sequence of ideals requires a sequence of dependent choices. A similar example from within set theory itself is

Proposition 7.7. Assume DC holds. Then a relation-class R is well-founded iff there are no infinite R-descending sequences, where an R-descending sequence is a function f with domain ω such that

$$(\forall n \in \omega)(\langle f(n^+), f(n)\rangle \in R).$$

Proof. If f is an R-descending sequence, then $\{f(n) \mid n \in \omega\}$ is a nonempty set with no R-minimal member. Conversely, if we are given a set a with no R-minimal member, then a straightforward application of DC produces an R-descending sequence of members of a. $\square$

In particular, in the presence of DC, the axiom of Foundation is equivalent to the assertion that there are no infinite $\in$-descending sequences.

There are also numerous applications of DC in topology; for example, in the proof that compactness implies sequential compactness in first countable spaces, and in the proof of Urysohn's Lemma about continuous real-valued functions on normal spaces.

Exercises

7.1.   A choice function $g: (\mathscr{P}a - \{\varnothing\}) \to a$ is said to be *orderly* if
$$g(b \cup c) = g(\{g(b), g(c)\})$$
for all nonempty sets $b, c \subseteq a$. Show that g is orderly iff it is induced (as in the first half of the proof of Proposition 7.2) by a well-ordering of a. [Hint: if g is so induced, we can recover the ordering by $(x \leqslant y) \Leftrightarrow (g(\{x, y\}) = x)$.]

7.2.   Let $g: (\mathscr{P}a - \{\varnothing\}) \to a$ be a choice function. We shall say that a subset c of $\mathscr{P}a$ is *closed* if
(i) $(b \in c$ and $b \neq a)$ implies $(b \cup \{g(a - b)\} \in c)$, and
(ii) the union of any chain of members of c belongs to c (so that in particular $\varnothing \in c$).
(a) Show that there is a unique smallest closed set $c_0 \subseteq \mathscr{P}a$.
(b) An element b of c_0 is called a *pinch point* if it is comparable (for the

inclusion ordering) with every other member of c_0. Show that the set of pinch points of c_0 is closed, and deduce that c_0 is linearly ordered by $\subset$.

(c) Show that for any $x \in a$ there is a unique largest $b \in c_0$ with $x \notin b$, and that for this b we have $x = g(a - b)$.

(d) Deduce that a is well-ordered by $<$, where

$$((x < y) \Leftrightarrow (\exists b \in c_0)((x \in b) \wedge (y \notin b))).$$

[This is (essentially) Zermelo's proof of the well-ordering theorem; cf. the Remark after Proposition 7.2.]

7.3. (a) Show that the axiom of choice is equivalent to the assertion that an arbitrary cartesian product of nonempty sets is nonempty.

(b) Let $(a_i \mid i \in I)$ be a family of nonempty sets, and for each $i \in I$ let $b_i = a_i \cup \{\infty\}$, where $\infty \notin \bigcup \{a_i \mid i \in I\}$. Topologize b_i so that its open sets are $\varnothing$, $\{\infty\}$ and all subsets with finite complements; observe that b_i is compact. By considering the closed subsets $c_i = \{f \in b \mid f(i) \neq \infty\}$ of the product space $b = \prod_{i \in I} b_i$, show that the assertion 'A product of compact topological spaces is compact' implies the axiom of choice.

7.4. A *selection function* on a set a is a function $f : \mathscr{P}a \to \mathscr{P}a$ such that $f(\varnothing) = \varnothing$ and $\varnothing \neq f(b) \subseteq b$ for all $b \in \mathscr{P}a - \{\varnothing\}$, the latter inclusion being strict unless b is a singleton. Suppose a has a selection function f, and let $g : \alpha \to 2$ ($\alpha \in \mathrm{ON}$) be any transfinite sequence of 0s and 1s. We define a subset $b(g)$ of a by recursion on α, as follows:

$$\begin{aligned}
&\text{if } \alpha = 0, && b(g) = a \\
&\text{if } \alpha = \beta^+ \text{ and } g(\beta) = 0, && b(g) = f(b(g|_\beta)) \\
&\text{if } \alpha = \beta^+ \text{ and } g(\beta) = 1, && b(g) = b(g|_\beta) - f(b(g|_\beta)) \\
&\text{if } \alpha \text{ is a nonzero limit,} && b(g) = \bigcap \{b(g|_\gamma) \mid \gamma < \alpha\}.
\end{aligned}$$

Show that there exists α such that $b(g)$ has at most one element for each g with domain α, and deduce that there exists an injection $a \to \mathscr{P}\alpha$. Conversely, if a is a subset of $\mathscr{P}\alpha$ for some ordinal α, show that a has a selection function. [Hint: for each $b \subseteq a$ with at least two elements, consider the least $\beta \in \alpha$ such that β belongs to some member of b but not to all of them.]

7.5. Let P be a set of propositional variables, and let S be a set of compound propositions with variables from P. Define a relation $\equiv$ on the free Boolean algebra FP by setting

$$p \equiv q \quad \text{iff} \quad S \vdash (p \Leftrightarrow q).$$

Show that $\equiv$ is a Boolean algebra congruence on FP (i.e. that the quotient $FP/\equiv$ inherits a Boolean algebra structure from FP), and that $FP/\equiv$ is nontrivial (i.e. has more than one element) iff S is consistent. Hence show that the Prime Ideal Theorem is equivalent to the Completeness Theorem for Propositional Logic.

7.6. Show that the Prime Ideal Theorem implies that every set can be linearly ordered. [Hint: given a set a, apply the Compactness Theorem to a propositional theory whose models are linear orderings of a.]

7.7. Show that the assertion 'Every set has a selection function' (cf. Exercise 7.4) implies that every set can be linearly ordered. [Hint: use Exercise 6.7(a).]

7.8. Show that the assertion 'Every set can be linearly ordered' implies that every family of nonempty finite sets has a choice function.

7.9. A *multiple-choice function* for a family of sets $(a_i \mid i \in I)$ is a function f with domain I such that, for each $i \in I$, $f(i)$ is a nonempty finite subset of a_i. If every family of nonempty sets has a multiple-choice function, prove that every linearly orderable set can be well-ordered.

7.10. Suppose that the power-set of any well-ordered set can be well-ordered. Show that V_α can be well-ordered for each $\alpha \in ON$, and deduce that AC holds. [This is harder than it looks; remember that we can't make infinitely many arbitrary choices of well-orderings for V_αs. To well-order V_α when α is a limit, begin by choosing a well-ordering of $\mathcal{P}\gamma$, where γ is the least ordinal not mapping injectively into V_α, and use this to construct a 'uniform' sequence of well-orderings of $(V_\beta \mid \beta < \alpha)$.]

7.11. Let K be a field, and let $(a_i \mid i \in I)$ be a family of nonempty sets. (We shall assume for convenience that the a_i are pairwise disjoint.) Let V be the K-vector space with basis $a = \bigcup \{a_i \mid i \in I\}$, and let V_0 be the subspace consisting of those finite sums $\sum \lambda_j x_j (\lambda_j \in K, x_j \in a)$ such that, for each $i \in I$, the sum of those coefficients λ_j for which $x_j \in a_i$ is 0. Suppose V_0 has a complementary subspace V_1. Show that for each $i \in I$ there is a unique $v_i \in V_1$ such that, for all $x \in a_i$, $(x - v_i) \in V_0$. Deduce that $(a_i \mid i \in I)$ has a multiple-choice function.

7.12. Use Zorn's Lemma to prove the axiom of Dependent Choices.

8

Cardinal arithmetic

Informally, cardinal arithmetic is what remains of set theory when you forget the fact that the members of a set are themselves sets (with their own particular structure), so that the only comparison you can make between two sets is to ask whether one has more members than the other. Slightly less informally, we'd like to define a *cardinal* (as Bertrand Russell proposed) to be an equivalence class of sets under the relation $\equiv$, where

$$(x \equiv y) \Leftrightarrow (\exists f)(f \text{ is a bijection } x \to y).$$

Unfortunately, the equivalence classes of this relation (except for the class $\{\varnothing\}$) are all proper classes; for example, the class of all singleton sets is itself in bijective correspondence with V. Since we wish to manipulate cardinals within set theory, it is convenient to find some way of representing these equivalence classes by sets; that is, we seek a function-class card: $V \to V$ with the property that

$$(\forall x, y)((\text{card}(x) = \text{card}(y)) \Leftrightarrow (x \equiv y)). \tag{$*$}$$

There are two standard ways of defining such a function, depending on whether or not we assume the axiom of Choice. If we do, then each $\equiv$-equivalence class contains at least one ordinal by Proposition 7.2, and so we may represent each class by the least ordinal which it contains; i.e.

$$\text{card}_1(x) = \bigcap \{\alpha \in \text{ON} \mid (x \equiv \alpha)\}.$$

[If you dislike forming the intersection of a class rather than a set, you may replace ON by the ordinal $\gamma(x)$ constructed in the proof of Lemma 7.1.] An ordinal is said to be *initial* if it is the least ordinal in some $\equiv$-equivalence class: every finite ordinal is initial, but infinite ordinals are much rarer. Note that the Hartogs function $x \mapsto \gamma(x)$ (cf. Lemma 7.1) takes values in the class of initial ordinals.

If we do not have the axiom of Choice, then it seems unlikely that we can define a cardinal function which picks out a single element of each equivalence class; but what we can do is to pick out a nonempty sub*set* of each class, by the following device (due to D. S. Scott). Define the *essential rank* of a set x [note: this is not standard terminology] by

$$\text{ess rank}(x) = \bigcap \{\alpha \in \text{ON} \mid (\exists y)((\text{rank}(y) = \alpha) \wedge (y \equiv x))\},$$

and then define

$$\text{card}_2(x) = \{ y \in V_{\text{ess rank}(x)+1} \mid (y \equiv x)\},$$

which is a set by Separation. Then it is clear form the definition that $x \equiv y$ implies $\text{ess rank}(x) = \text{ess rank}(y)$ and hence $\text{card}_2(x) = \text{card}_2(y)$; and conversely since we have

$$(\forall x)((\text{card}_2(x) \neq \varnothing) \wedge (\forall z \in \text{card}_2(x))(z \equiv x))$$

we can deduce $x \equiv y$ from $\text{card}_2(x) = \text{card}_2(y)$.

Henceforth, we shall use 'card' to denote either of the functions card_1 and card_2; but it really doesn't matter which one we use, since the only property of the cardinal function we shall ever require is the property (∗). [However, even if we are using card_1, we shall find it convenient to distinguish notationally between an infinite initial ordinal and its cardinal, if only because the sum and product operations we're going to introduce on cardinals don't agree with ordinal sum and product. On the other hand, we *shall* tend to confuse finite cardinals with natural numbers, even if we are using card_2. Note also that we shall tend to transfer epithets from sets to their cardinals; thus we shall speak of a cardinal being infinite, well-orderable, etc., when we mean that the corresponding property holds for the sets whose cardinal it is.] Thus when we define arithmetic operations, etc., on cardinals, what we are really doing is defining operations on sets and verifying that they respect the relation $\equiv$.

We shall find it convenient, for this chapter, to adopt the convention that $m, n, p, \ldots$ will denote cardinals and the corresponding capital letters $M, N, P, \ldots$ denote sets of the corresponding cardinalities. We also define a normal function (cf.

Exercise 6.9) $\alpha \mapsto \omega_\alpha$ by recursion:

$$\omega_0 = \omega$$

$$\omega_\alpha = \gamma(\omega_\beta) \qquad \text{if } \alpha = \beta^+$$

$$\omega_\alpha = \bigcup \{\omega_\beta \mid \beta \in \alpha\} \quad \text{if } \alpha \text{ is a nonzero limit.}$$

Then it is easy to see that the ω_α are all the infinite initial ordinals. Following Cantor, we shall write $\aleph_\alpha$ for $\text{card}(\omega_\alpha)$; in particular, $\aleph_0$ is the cardinality of all countably infinite sets.

With these preliminaries out of the way, we may start doing some cardinal arithmetic. If m and n are cardinals, we write $m \leqslant n$ to mean that there exists an injection $M \to N$, and $m \leqslant^* n$ to mean that either $m = 0$ ($= \text{card}(\varnothing)$) or there exists a surjection $N \to M$. Clearly, these definitions do not depend on the choice of representatives M and N. (We also write $m < n$ for ($m \leqslant n$ and $m \neq n$), etc.) They are related by

Lemma 8.1. (i) If $m \leqslant n$, then $m \leqslant^* n$.

(ii) If $m \leqslant^* n$ and n is well-orderable, then $m \leqslant n$.

Proof. (i) The case $m = 0$ is trivial, so suppose $M \neq \varnothing$ and we have an injection $f: M \to N$. Choose $x_0 \in M$ and define $g: N \to M$ by

$$g(y) = \text{the unique } x \in M \text{ with } f(x) = y, \text{ if such exists,}$$

$$= x_0 \text{ otherwise.}$$

Then g is a surjection.

(ii) Again, the case $m = 0$ is trivial. Suppose we have a surjection $g: N \to M$ and a well-ordering $<$ of N. Then we may define an injection $f: M \to N$ sending $x \in M$ to the $<$-least element of $\{y \in N \mid g(y) = x\}$. $\square$

In particular, if AC holds then $\leqslant$ and $\leqslant^*$ coincide; in general, however, they may be different. It is easy to see that both $\leqslant$ and $\leqslant^*$ are reflexive and transitive, and $\leqslant$ is actually a partial order:

Proposition 8.2 (the Cantor–Bernstein Theorem). If $m \leqslant n$ and $n \leqslant m$, then $m = n$.

Proof. Let $f: M \to N$ and $g: N \to M$ be injections. To construct a bijection $M \to N$, it is sufficient to partition M as a disjoint union $M_0 \cup M_1$, and N as $N_0 \cup N_1$, such that f restricts to a bijection $M_0 \to N_0$ and g to a bijection $N_1 \to M_1$. But finding such a pair of partitions is equivalent to finding a fixed point of the function

$h: \mathscr{P}M \to \mathscr{P}M$ defined by $h(A) = M - g[N - f[A]]$ (where we write $f[A]$ for $\{f(x) \mid x \in A\}$, etc.). It is clear that the function h preserves inclusions, i.e. $A_1 \subseteq A_2$ implies $h(A_1) \subseteq h(A_2)$.

Let $X = \{A \in \mathscr{P}M \mid A \subseteq h(A)\}$, and let $M_0 = \bigcup X$. Then for every $A \in X$, we have $A \subseteq h(A) \subseteq h(M_0)$, and hence $M_0 \subseteq h(M_0)$. Hence also $h(M_0) \subseteq h(h(M_0))$, so $h(M_0) \in X$ and thus $h(M_0) \subseteq M_0$. So M_0 is a fixed point of h, as required. $\square$

However, $\leqslant^*$ may fail to be a partial order if we do not assume AC (see Exercise 8.6). The next obvious question to ask is when $<$ is a linear order.

Proposition 8.3. In a model of set theory, the following are equivalent:

   (i) $<$ is a linear ordering of the class of cardinals.
   (ii) $<$ is a well-ordering of the class of cardinals.
   (iii) AC holds.

Proof. (iii) $\Rightarrow$ (ii): If AC holds, then the class of cardinals is order-isomorphic to the class of initial ordinals, which is well-ordered as a subclass of ON.

(ii) $\Rightarrow$ (i) is trivial.

(i) $\Rightarrow$ (iii): If x is any set, then by Lemma 7.1 we have $\text{card}(\gamma(x)) \not\leqslant \text{card}(x)$. So (i) implies that we have an injection $x \to \gamma(x)$, which enables us to well-order x. $\square$

In the absence of AC, the law of trichotomy holds for finite cardinals, but it may even break down when one of the cardinals involved is $\aleph_0$. If $m \not\leqslant \aleph_0$, then an easy induction shows that we have injections $n \to M$ for each natural number n, but it requires an application of (countable) choice to construct an injection $\omega \to M$. A set M is said to be *Dedekind infinite* if $\aleph_0 \leqslant \text{card}(M)$; the following equivalence is easily established, even without Choice.

Proposition 8.4. A set M is Dedekind infinite iff there exists an injection $M \to M$ which is not surjective.

Proof. Given an injection $f: \omega \to M$, define $g: M \to M$ by

$$g(x) = f(n + 1) \quad \text{if } (\exists n \in \omega)(f(n) = x)$$
$$= x \qquad \qquad \text{otherwise.}$$

Then g is injective, but not surjective since $f(0) \notin \operatorname{im}(g)$. Conversely, given an injection $g: M \to M$ and $x \in M - \operatorname{im}(g)$, define $f: \omega \to M$ recursively by $f(0) = x$, $f(n + 1) = g(f(n))$; then an easy induction shows that f is injective. □

Next we define cardinal addition, multiplication and exponentiation:

$m + n = \operatorname{card}(M \amalg N)$ $(= \operatorname{card}(M \cup N)$ if M and N are disjoint)

$m \cdot n = \operatorname{card}(M \times N)$

$m^n = \operatorname{card}(M^N)$ (where $M^N = \{f \mid f: N \to M\}$).

If we assume AC, we may also define infinite sums and products of cardinals by $\sum_{i \in I} m_i = \operatorname{card}(\mathscr{P}_{i \in I} M_i)$ (where the infinite disjoint union $\coprod_{i \in I} M_i$ is defined to be $\bigcup \{M_i \times \{i\} \mid i \in I\}$) and $\prod_{i \in I} m_i = \operatorname{card}(\prod_{i \in I} M_i)$. However, these operations may not be well-defined without AC, since if we are merely given that $\operatorname{card}(M_i) = \operatorname{card}(M'_i)$ for each i, we cannot prove $\operatorname{card}(\coprod_{i \in I} M_i) = \operatorname{card}(\coprod_{i \in I} M'_i)$ without choosing a particular bijection $M_i \to M'_i$ for each i.

The next lemma summarizes the properties of the finitary cardinal operations (we leave those of the infinitary operations as an exercise):

Lemma 8.5. (a) If $m \leqslant n$, then $m + p \leqslant n + p$, $m \cdot p \leqslant n \cdot p$ and $m^p \leqslant n^p$.

(b) If $m \leqslant^* n$, then $m + p \leqslant^* n + p$, $m \cdot p \leqslant^* n \cdot p$ and $p^m \leqslant p^n$ unless $p = m = 0$.

(c) $0 + m = m, 0 \cdot m = 0, 1 \cdot m = m, m^0 = 1, m^1 = m$ and $1^m = 1$.

(d) $1 + m = m$ iff m is Dedekind infinite.

(e) $m + n = n + m$ and $m + (n + p) = (m + n) + p$.

(f) $m \cdot n = n \cdot m$ and $m \cdot (n \cdot p) = (m \cdot n) \cdot p$.

(g) $m \cdot (n + p) = m \cdot n + m \cdot p$ and $m^{(n+p)} = m^n \cdot m^p$.

(h) $m^{(n \cdot p)} = (m^n)^p$ and $(m \cdot n)^p = m^p \cdot n^p$.

(j) If α and β are ordinals, then $\operatorname{card}(\alpha + \beta) = \operatorname{card}(\alpha) + \operatorname{card}(\beta)$ and $\operatorname{card}(\alpha \cdot \beta) = \operatorname{card}(\alpha) \cdot \operatorname{card}(\beta)$, but $\operatorname{card}(\alpha^\beta) \neq \operatorname{card}(\alpha)^{\operatorname{card}(\beta)}$ in general.

(k) If $m \geqslant 2$ and $n \geqslant 2$, then $m + n \leqslant m \cdot n$.

Proof. Almost all of these are trivial when translated into assertions about bijections (etc.) between certain sets. We comment on a few of them:

(b)(iii): Given a surjection $f: N \to M$, the operation of composing with f defines a function $P^M \to P^N$, which is easily seen to be injective. If $m = 0$, then $p^m = 1 \leqslant p^n$ unless $p = 0$.

(d) follows from Proposition 8.4: given an injection $g: M \to M$ as in the proof of 8.4, we have $m = m + \mathrm{card}(M - \mathrm{im}(g)) \geqslant m + 1 \geqslant m$, so $m = m + 1$ by Proposition 8.2. The converse is similar.

(g)(ii): We have a bijection $M^{(N \cup P)} \to M^N \times M^P$ (if N and P are disjoint) sending f to $\langle f|_N, f|_P \rangle$.

(j)(i) and (ii) follow from the synthetic definitions of ordinal sum and product. For (j)(iii) take $\alpha = 2$, $\beta = \omega$; then $\mathrm{card}(\alpha^\beta) = \aleph_0$ but $\mathrm{card}(\alpha)^{\mathrm{card}(\beta)} = 2^{\aleph_0} = \mathrm{card}(\mathscr{P}\omega) \neq \aleph_0$ by Cantor's diagonal argument.

(k) Choose distinct elements $x_0, x_1 \in M$ and $y_0, y_1 \in N$. Then define $f: M \amalg N \to M \times N$ by

$$f(\langle x, 0 \rangle) = \langle x, y_0 \rangle \quad (x \in M)$$
$$f(\langle y, 1 \rangle) = \langle x_0, y \rangle \quad (y \in N, \, y \neq y_0)$$
$$f(\langle y_0, 1 \rangle) = \langle x_1, y_1 \rangle.$$

It is clear that f is injective. $\square$

In the presence of AC, the addition and multiplication of infinite cardinals becomes very simple; recall that in this case any infinite cardinal is of the form $\aleph_\alpha$ for some ordinal α.

Lemma 8.6. For any α, $\aleph_\alpha \cdot \aleph_\alpha = \aleph_\alpha$.

Proof by induction on α. Define a well-ordering of $\omega_\alpha \times \omega_\alpha$ by setting

$$(\langle \beta_1, \gamma_1 \rangle < \langle \beta_2, \gamma_2 \rangle) \Leftrightarrow (((\beta_1 \cup \gamma_1) < (\beta_2 \cup \gamma_2))$$
$$\vee (((\beta_1 \cup \gamma_1) = (\beta_2 \cup \gamma_2)) \wedge (\beta_1 < \beta_2))$$
$$\vee (\gamma_1 < \gamma_2 \leqslant \beta_1 = \beta_2)).$$

It is straightforward to verify that this is indeed a well-ordering; moreover, for each $\delta < \omega_\alpha$, the subset $\delta \times \delta$ is an initial segment of $\omega_\alpha \times \omega_\alpha$, and every proper initial segment of $\omega_\alpha \times \omega_\alpha$ is contained in one of this form. But for any such δ, either δ is finite (in which case $\mathrm{card}(\delta \times \delta) < \aleph_0 \leqslant \aleph_\alpha$) or by the inductive hypothesis we have $\mathrm{card}(\delta \times \delta) = \mathrm{card}(\delta) < \aleph_\alpha$. So we have constructed a well-ordering of $\omega_\alpha \times \omega_\alpha$ in which every proper initial segment has cardinality $< \aleph_\alpha$, which means that its order-type is at most ω_α. But $\mathrm{card}(\omega_\alpha \times \omega_\alpha) \geqslant \aleph_\alpha$ since $\aleph_\alpha \geqslant 1$, and so the order-type must be exactly ω_α; i.e. it defines a bijection $\omega_\alpha \times \omega_\alpha \to \omega_\alpha$. $\square$

Corollary 8.7. For any α, $\aleph_\alpha + \aleph_\alpha = \aleph_\alpha$.

Proof. From Lemma 8.5, we have inequalities $\aleph_\alpha \leqslant \aleph_\alpha + \aleph_\alpha \leqslant \aleph_\alpha \cdot \aleph_\alpha$, so the result follows from Lemma 8.6 and Proposition 8.2. Alternatively, it can be proved directly by constructing a suitable well-ordering of $\omega_\alpha \amalg \omega_\alpha$ and arguing as in the proof of Lemma 8.6 (cf. Exercise 8.7). $\square$

There is in fact a converse to Lemma 8.6: if the equality $m^2 = m$ holds for all infinite cardinals m, then AC holds (see Exercise 8.3). On the other hand, the assertion '$2m = m$ for all infinite m', though not provable from the axioms of ZF, does not imply AC.

Corollary 8.8. For all α and β, $\aleph_\alpha + \aleph_\beta = \aleph_\alpha \cdot \aleph_\beta = \aleph_{(\alpha \cup \beta)}$.

Proof. Suppose $\alpha \geqslant \beta$. Then we have

$$\aleph_\alpha \leqslant \aleph_\alpha + \aleph_\beta \leqslant \aleph_\alpha \cdot \aleph_\beta \leqslant \aleph_\alpha \cdot \aleph_\alpha = \aleph_\alpha$$

by Lemmas 8.5 and 8.6; an application of Proposition 8.2 yields the result. $\square$

There are similar results for infinite sums of infinite cardinals, if we assume AC: if $\text{card}(I) \leqslant \aleph_\alpha$ and $m_i \leqslant \aleph_\alpha$ for each $i \in I$, then $\sum_{i \in I} m_i \leqslant \aleph_\alpha$, since we can use AC to construct an injection $\coprod_{i \in I} M_i \to \omega_\alpha \times \omega_\alpha$. (The case $\alpha = 0$ of this is the familiar assertion that a countable union of countable sets is countable; note that it requires (countable) Choice even in this case.) To get results on infinite products, however, we need to investigate the behaviour of cardinal exponentials; and this is less straightforward than addition or multiplication, even with AC.

We can use Proposition 8.2 to achieve some reductions in the problem of determining cardinal exponentials: for example, for any $\beta \leqslant \alpha$ we have

$$2^{\aleph_\alpha} \leqslant \aleph_\beta^{\aleph_\alpha} \leqslant (2^{\aleph_\alpha})^{\aleph_\alpha} = 2^{(\aleph_\alpha \cdot \aleph_\alpha)} = 2^{\aleph_\alpha},$$

and so our attention is focused on the function $f : \text{ON} \to \text{ON}$ defined by $2^{\aleph_\alpha} = \aleph_{f(\alpha)}$. Cantor's diagonal argument tells us that $\alpha < f(\alpha)$ for all α; and it is clear that $\alpha \leqslant \beta$ implies $f(\alpha) \leqslant f(\beta)$. Cantor spent many years trying to prove the *Continuum Hypothesis* (CH) that (in this notation) $f(0) = 1$; i.e. that the cardinality $2^{\aleph_0}$ of the continuum (the set of all real numbers) is the first uncountable (well-orderable) cardinal. A natural extension of this was the *Generalized Continuum*

Hypothesis (GCH) that $f(\alpha) = \alpha + 1$ for all α. In 1938 K. Gödel showed that the GCH is consistent relative to the other axioms of set theory; it does, however, imply AC (this follows from the result of Exercise 7.10, or from Exercise 8.5 if GCH is reformulated so that it doesn't refer to alephs). In 1963 P. Cohen showed that it is consistent relative to the other axioms of set theory (including AC) that CH doesn't hold; in fact, using his methods, one can show that there are virtually no restrictions on the function f apart from those we have mentioned, plus certain conditions involving limit ordinals (cf. Exercises 8.8 and 8.9).

Exercises

[Note: the axiom of Choice should not be assumed in Exercises 8.1–8.6, but is required in Exercises 8.8 and 8.9.]

8.1. If $m \not\leqslant \aleph_0$, show that $\aleph_0 \leqslant *2^m$. [Hint: consider the mapping which sends a subset of M to its cardinality if it's finite, and to 0 otherwise.] Deduce that any cardinal of the form 2^{2^m} is either $< \aleph_0$ or $> \aleph_0$.

8.2. If $m + n = m \cdot n$, show that either $n \leqslant m$ or $m \leqslant *n$. [Hint: given a bijection $f : M \amalg N \to M \times N$, consider whether the composite

$$N \xrightarrow{} M \amalg N \xrightarrow{f} M \times N \xrightarrow{\pi} M$$

is surjective, where π denotes projection on the first factor.]

8.3. Let $m = \mathrm{card}(M)$ be an infinite cardinal, and let $n = \mathrm{card}(\gamma(M))$. If $(m + n)^2 = m + n$, deduce from Exercise 8.2 that $m \leqslant *n$, and hence show that M can be well-ordered. Deduce that the assertion 'For all Dedekind infinite cardinals m, $m^2 = m$' is equivalent to the axiom of Choice.

8.4. We write $m \lhd n$ to mean that $m < n$ and, for all p, $m \leqslant p \leqslant n$ implies either $m = p$ or $p = n$. If $m \lhd n$ and n is Dedekind infinite, show that either $\gamma(M) = \gamma(N)$ or $n = n + \mathrm{card}(\gamma(M)) = m + \mathrm{card}(\gamma(M))$.

8.5. Show that $\mathrm{card}(\gamma(M)) \leqslant *2^{(m^2)}$ [hint: consider the proof of Lemma 7.1]. Now suppose $m \lhd 2^m$ for all infinite m; let m be an infinite cardinal, and define a sequence of cardinals m_i by

$$m_0 = m, \quad m_{i+1} = 2^{m_i}.$$

Using Exercises 8.1 and 8.4, show that for some $i \in \omega$ we have $m_i^2 = m_i$ and $m_{i+1} = m_i + \mathrm{card}(\gamma(M_i))$. Deduce, using Exercise 8.3, that M can be well-ordered. [Thus the GCH, in the form '$m \lhd 2^m$ for all infinite m', implies the axiom of Choice.]

8.6. Let us say that two elements R, S of $\mathscr{P}(\omega \times \omega)$ are *similar* if there exists a permutation f of ω such that

$$(\forall x, y \in \omega)((\langle x, y \rangle \in R) \Leftrightarrow (\langle f(x), f(y) \rangle \in S)).$$

By considering equivalence relations on ω with finite equivalence classes, or otherwise, show that there is a set of pairwise dissimilar elements of $\mathscr{P}(\omega \times \omega)$ having cardinality $2^{\aleph_0}$. By considering similarity classes of well-orderings, show that the set of all similarity classes of elements of $\mathscr{P}(\omega \times \omega)$ has cardinality $\geq \aleph_1 + 2^{\aleph_0}$. Deduce that if every well-orderable set of real numbers is countable, then there exist cardinals m and n with $m \leqslant n$ and $n \leqslant^* m$ but $m \neq n$.

8.7. Define a well-ordering of $\omega_\alpha \amalg \omega_\alpha$ having order-type ω_α.

8.8. Let $(m_i \mid i \in I)$ and $(n_i \mid i \in I)$ be two families of cardinals, and suppose that for each $i \in I$ we have $m_i < n_i$. Prove that

$$\sum_{i \in I} m_i < \prod_{i \in I} n_i.$$

What can you deduce from this result (a) when $m_i = 0$ for all i, and (b) when $m_i = 1$ and $n_i = 2$ for all i? By taking $I = \omega, m_i = \aleph_i$ and $n_i = \aleph_\omega$ for all i, show that $2^{\aleph_0} \neq \aleph_\omega$.

8.9. Show that

$$2^{(\sum_{i \in I} m_i)} = \prod_{i \in I} (2^{m_i})$$

for any family of cardinals $(m_i \mid i \in I)$. Deduce that if $2^{\aleph_i} = \aleph_{i+1}$ for all $i \in \omega$, then $2^{\aleph_\omega} = (\aleph_\omega)^{\aleph_0}$.

9

Consistency and independence

Throughout our discussion of ZF set theory, we have been tacitly assuming that it is consistent – or equivalently (if we assume the Completeness Theorem holds in our meta-universe) that it has a model. If so, then we know from the Löwenheim–Skolem theorems that this model will not be unique, even up to isomorphism; but we might still hope that ZF is a complete theory in the sense defined in the proof of Theorem 3.7, i.e. that every sentence in the language of ZF is either provable or refutable from the axioms. (Again assuming the Completeness Theorem, this is equivalent to saying that any two models of ZF are *elementarily equivalent*, i.e. that they satisfy exactly the same sentences.)

It was thus a natural programme, once the axioms of ZF set theory had been established, to seek to prove that it was both consistent and complete (or alternatively, if it turned out not to be complete, to seek extra axioms which would make it into a complete theory). This programme was first explicitly proposed by D. Hilbert, around 1920; but just ten years later K. Gödel showed that it could not be carried out, by proving his two Incompleteness Theorems. These assert, respectively, that if set theory is consistent then it is incomplete, and that if set theory is consistent then no proof of its consistency can be formalized within set theory. Moreover, the incompleteness of ZF is an 'essential' one, which cannot be remedied by adding finitely many new axioms or axiom-schemes. (In contrast, the theory of algebraically closed fields is incomplete, since the sentence '$1 + 1 = 0$' is satisfied in some algebraically closed fields but not in others; but if we add an extra axiom or scheme of axioms to specify what the characteristic is, we do in fact obtain a complete theory.)

Before giving the precise statements of the Incompleteness Theorems, we need a couple of preliminaries: the notion of an interpretation of one theory in another, and the idea of formalizing proofs and provability within Peano arithmetic. For the first of these, we begin with an example: recall that we have already observed that, if V is any model of set theory, then the class of natural numbers in V has the structure of a model of Peano arithmetic. Thus we have a recipe for constructing models of PA from models of ZF; and the recipe does not depend on any particular features of the ZF-model from which we start, because it can be described entirely in terms of the syntax of the two theories.

More generally, if T_1 and T_2 are first-order theories in languages $\mathscr{L}_1$ and $\mathscr{L}_2$, an *interpretation* $v: T_1 \to T_2$ is a 'uniform' recipe for constructing T_1-models from T_2-models. It is specified by giving, first, a formula $v(x)$ of $\mathscr{L}_2$ (the idea being that if M is our T_2-model then $[v(x)]_M$ will be the underlying set of our T_1-model; for convenience we shall assume that v has just one free variable x, although it is possible to consider more general interpretations), and then for each primitive function-symbol ω (respectively predicate-symbol ϕ) of $\mathscr{L}_1$ a term v_ω (respectively a formula v_ϕ) of $\mathscr{L}_2$ with the same number of free variables. Given these, we may then inductively assign to each term t or formula p of $\mathscr{L}_1$ a corresponding term v_t or formula v_p of $\mathscr{L}_2$, the only nontrivial point in the induction being that quantifiers must be 'relativized to $[v(x)]$', i.e. $v_{(\forall x)p}$ is $(\forall x)(v(x) \Rightarrow v_p)$. Then the assertion that v is an interpretation of T_1 in T_2 means that, for each axiom p of T_1, v_p is a theorem of T_2.

Returning to our example of an interpretation PA $\to$ ZF, we have for $v(x)$ the formula 'x is a natural number' (i.e.

$$(\forall y)((y \text{ is a successor set}) \Rightarrow (x \in y)) \quad),$$

the constant 0 and the unary operation s of the language of PA are interpreted by the constant $\varnothing$ and the unary operation $(\ )^+$, and the binary operations a and m are interpreted by ordinal addition and multiplication. [We are cheating slightly here, in that our original language for ZF did not have any primitive function-symbols, and so did not contain the terms mentioned above; if we wish to interpret PA in ZF as formulated in this language, then we must reformulate PA in a language without primitive function-symbols, along the lines indicated in Exercise 3.5.]

For the second ingredient of the Incompleteness Theorems, we need to recall the notion of a recursively presented theory, which we last met (in the particular case of an algebraic theory) in Chapter 4. For the general case, we suppose first that we have a language $\mathscr{L}$ with countably many primitive symbols, which are enumerated in such a way that the function assigning their arities is recursive. [Readers who have skipped Chapter 4 will lose relatively little, in the discussion which follows, by substituting the phrase 'definable by a formula in the language of PA' whenever they see the word 'recursive'; those who have read Chapter 4 should note that we shall be relying heavily on Theorem 4.13, which proved that every recursive function is PA-definable.] Note that if (as is the case with both ZF and PA) the set of primitive symbols of the language is actually finite, then the recursiveness of the arity function imposes no further restriction.

We further assume that we are given an enumeration, not only of the primitive symbols of our particular language $\mathscr{L}$, but also of the countably many 'standard symbols' (variables, parentheses, logical connectives, etc.) which are used in constructing terms and formulae of $\mathscr{L}$. A term t or formula p may thus be specified by giving the finite sequence $(n_0, n_1, \ldots, n_k)$ of natural numbers corresponding to the individual symbols which make it up; and this sequence may be specified by the single number

$$2^{n_0} 3^{n_1} \ldots p_k^{n_k}$$

(where p_k denotes the $(k + 1)$th prime number), which we call the *code* or *Gödel number* of the term or formula, and denote by $\ulcorner t \urcorner$ or $\ulcorner p \urcorner$ as appropriate. A straightforward extension of Remark 1.1 shows that we can construct an algorithm for determining whether a given number is the code for a formula of $\mathscr{L}$; i.e. the set of such codes is recursive.

The precise details of the coding whereby we represent formulae of $\mathscr{L}$ by natural numbers are not of great importance; but it is important to note that, for any reasonable coding, the (algorithmic) operations which we are accustomed to carry out on terms and formulae will be coded by recursive functions. For example, for each variable x there will be a recursive function sub_x such that

$$\text{sub}_x(\ulcorner p \urcorner, \ulcorner t \urcorner) = \ulcorner p[t/x] \urcorner$$

for each formula p and term t of $\mathscr{L}$. We now say that a theory T, in a language $\mathscr{L}$ as above, is *recursively presented* if the set of codes for axioms of T is recursive. Clearly, this will be the case if T is specified by a finite number of individual axioms plus a finite number of axiom-schemes; thus both PA and ZF are recursively presented.

Now a derivation in $\mathscr{L}$ is just a finite string of formulae, and so we can also code them by natural numbers in a straightforward way. Moreover, for a recursively presented theory T, the problem of determining whether a given finite string of formulae is a derivation from the axioms of T is algorithmically soluble; thus we have a formula $\mathrm{Der}_T(x, y)$ in the language of PA, such that $\mathrm{PA} \vdash \mathrm{Der}_T(m, n)$ (for natural numbers m and n, regarded as closed terms in the language of PA) iff m is the code of a derivation from the axioms of T of the formula whose code is n.

When we are dealing with recursively presented theories, it makes sense to consider interpretations $v: T_1 \to T_2$ which are *recursive*, in the following sense: first, the functions $\ulcorner t \urcorner \mapsto \ulcorner v_t \urcorner$ and $\ulcorner p \urcorner \mapsto \ulcorner v_p \urcorner$ are recursive (for this, it clearly suffices that their restrictions to the primitive symbols of $\mathscr{L}_1$ should be recursive), and second, there is a recursive function which, given the code for an axiom p of T_1, produces the code for a derivation of v_p from the axioms of T_2. Given this, the process of converting a derivation of an arbitrary formula p from the axioms of T_1 into a derivation of v_p from the axioms of T_2 becomes algorithmic; so there are recursive functions f and g such that $f(\ulcorner p \urcorner) = \ulcorner v_p \urcorner$ and

$$\mathrm{PA} \vdash (\forall x, y)(\mathrm{Der}_{T_1}(x, y) \Rightarrow \mathrm{Der}_{T_2}(g(x), f(y))).$$

Until further notice, we shall assume that T is a recursively presented theory equipped with a recursive interpretation $v: \mathrm{PA} \to T$. (For example, T might be PA itself, or ZF, or any extension of either of these theories obtained by adding finitely many new axioms or axiom-schemes.) To simplify the notation, we shall identify formulae of PA with their interpretations in the language of T, whenever it is possible to do so without confusion. We now introduce the unary predicate 'is provable in T' by

$$\mathrm{Pr}_T(y) \Leftrightarrow (\exists x)(\mathrm{Der}_T(x, y)).$$

Clearly, if $T \vdash p$ then $\mathrm{PA} \vdash \mathrm{Pr}_T(\ulcorner p \urcorner)$; the converse may not be true, because PA does not have witnesses (i.e. the fact that we can prove a sentence of the form $(\exists x)q(x)$ in PA does not imply the existence of a closed term m such that $PA \vdash q(m)$). We can, however, formalize the above implication in the sense that we can prove

$$PA \vdash (Pr_T(\ulcorner p \urcorner) \Rightarrow Pr_T(\ulcorner Pr_T(\ulcorner p \urcorner) \urcorner)),$$

because the process of constructing a proof of $Pr_T(\ulcorner p \urcorner)$ in PA from a proof of p in T is an algorithmic one, and so can be coded by a recursive function. For similar reasons, 'formalized modus ponens' is provable in PA, i.e.

$$PA \vdash (\forall x, y)((Pr_T(x) \wedge Pr_T(\mathrm{imp}(x, y))) \Rightarrow Pr_T(y)),$$

where imp is a recursive function such that $\mathrm{imp}(\ulcorner p \urcorner, \ulcorner q \urcorner) = \ulcorner (p \Rightarrow q) \urcorner$.

Now let $p(x)$ be the formula $\neg Pr_T(x)$ of PA (or its interpretation in the language of T). Let c be a recursive function such that $c(n) = \ulcorner n \urcorner$ for each natural number n (where the 'n' on the right-hand side is regarded as a closed term in the language of T), and let $q(y)$ be the formula $p(\mathrm{sub}_x(y, c(y)))$. Finally, let m be the code for $q(x)$. Then the sentence $q(m)$ asserts its own unprovability in T, in that we have

$$q(m) = p(\mathrm{sub}_x(m, c(m))) = p(\mathrm{sub}_x(\ulcorner q(x) \urcorner, \ulcorner m \urcorner))$$
$$= p(\ulcorner q(m) \urcorner) = \neg Pr_T(\ulcorner q(m) \urcorner).$$

Clearly, if T is consistent then we cannot have $T \vdash q(m)$; for, if we did, then we should have $PA \vdash Pr_T(\ulcorner q(m) \urcorner)$ and hence $T \vdash \neg q(m)$. Thus $q(m)$ is true (in an informal sense) but not provable in T; this gives us an informal version of the First Incompleteness Theorem.

However, it is possible to have a consistent theory T such that $T \vdash \neg q(m)$, because of the lack of witnesses in PA which we mentioned earlier. Thus, to get a formal incompleteness proof, we need to strengthen the assumption of consistency in some way. There are various ways of doing this: the one originally used by Gödel was to demand that T should be ω-consistent, which means that if $p(x)$ is a formula of PA with one free variable x, then we do not simultaneously have $T \vdash (\exists x)p(x)$ and $T \vdash \neg p(n)$ for each closed term n. If this condition holds, then it is easy to see that we cannot have $T \vdash \neg q(m)$; thus we have established

Theorem 9.1 (First Incompleteness Theorem). Let T be a recursively presented, ω-consistent first-order theory containing a recursive interpretation of PA. Then T is incomplete. $\square$

For the Second Incompleteness Theorem, let $\text{Con}_T = \neg\text{Pr}_T(\ulcorner\bot\urcorner)$ be the formalized version of the assertion that T is consistent.

Theorem 9.2 (Second Incompleteness Theorem). Let T be a recursively presented, consistent first-order theory containing a recursive interpretation of PA. Then $T \not\vdash \text{Con}_T$.

Proof. Let $q(m)$ be the unprovable sentence constructed above. We shall show that $\text{PA} \vdash (q(m) \Leftrightarrow \text{Con}_T)$, so that Con_T is also unprovable in T.

In one direction, the implication is easy: $(\bot \Rightarrow q(m))$ is a theorem of T (as of any first-order theory; cf. Exercise 2.2), and so $\text{Pr}_T(\ulcorner(\bot \Rightarrow q(m))\urcorner)$ is a theorem of PA. By formalized modus ponens, we obtain

$$\text{PA} \vdash (\text{Pr}_T(\ulcorner\bot\urcorner) \Rightarrow \text{Pr}_T(\ulcorner q(m)\urcorner)).$$

But $q(m)$ is the sentence $\neg\text{Pr}_T(\ulcorner q(m)\urcorner)$, so we deduce

$$\text{PA} \vdash (q(m) \Rightarrow \neg\text{Pr}_T(\ulcorner\bot\urcorner)).$$

Conversely, we have $\text{PA} \vdash (\text{Pr}_T(\ulcorner q(m)\urcorner) \Rightarrow \text{Pr}_T(\ulcorner\text{Pr}_T(\ulcorner q(m)\urcorner)\urcorner))$, which is equivalent to $\text{PA} \vdash (\text{Pr}_T(\ulcorner q(m)\urcorner) \Rightarrow \text{Pr}_T(\ulcorner\neg q(m)\urcorner))$. An elementary logical argument, formalized in PA, then yields

$$\text{PA} \vdash (\text{Pr}_T(\ulcorner q(m)\urcorner) \Rightarrow \text{Pr}_T(\ulcorner(q(m) \wedge \neg q(m))\urcorner)),$$

whence $\text{PA} \vdash (\text{Pr}_T(\ulcorner q(m)\urcorner) \Rightarrow \text{Pr}_T(\ulcorner\bot\urcorner))$. So

$$\text{PA} \vdash (\text{Con}_T \Rightarrow q(m)). \qquad \square$$

The message of Theorem 9.2 is that we cannot hope to give a 'self-supporting' proof of the consistency of any formal system which is powerful enough to be used as a foundation for mathematics; in order to prove the consistency of such a system T, we have to make assumptions which are stronger than T. [There is an obvious minimal such assumption: namely, adopt Con_T as an extra axiom. However, for obvious reasons, we cannot hope to deduce the consistency of the theory $T \cup \{\text{Con}_T\}$ from that of T.] On the other hand, we can hope to prove *relative* consistency results of the form $(\text{Con}_{T_1} \Rightarrow \text{Con}_{T_2})$, where T_1 and T_2 are as in Theorem 9.2; the rest of this chapter will be devoted to results of this general kind. In particular, we shall be interested in results of the form

$$(\text{Con}_{\text{ZF}} \Rightarrow \text{Con}_{\text{ZF} - \{p\} \cup \{\neg p\}})$$

where p is one of the axioms of ZF; we call this an *independence* result for the axiom p.

The obvious technique for proving the consistency of T_2 relative to T_1 is to construct a recursive interpretation $v: T_2 \to T_1$; for we then have

$$\text{PA} \vdash (\text{Pr}_{T_2}(\ulcorner p \urcorner) \Rightarrow \text{Pr}_{T_1}(\ulcorner v_p \urcorner))$$

for any p in the language of T_2, and taking $p = \bot$ yields the desired result. In practice, it is almost always easier to think of these interpretations in semantic terms, as constructions on models of T_1; but it is important to notice that the constructions we describe operate 'uniformly' on all models of T_1, and that they are really the reflections of syntactic interpretations as defined earlier (so that, in particular, the relative consistency results we obtain do not depend on the Completeness Theorem).

Let us, for the moment, take T_1 to be ZF. A particularly simple class of interpretations $T_2 \to$ ZF arises from what are commonly called *standard models*: in these we take some class M to be the underlying (meta-)set of our T_2-model, and interpret the membership predicate in the language of T_2 by itself in the language of ZF. (The term *inner model* is also used if the class M is a set.) Many of the axioms of ZF are inherited in a straightforward way by substructures of this kind, as we saw in the Exercises at the end of Chapter 5: in particular, if M is transitive then it inherits Extensionality and Foundation from V, and if M is 'super-transitive' (i.e. $x \subseteq y \in M$ implies $x \in M$) then it inherits Separation. (However, super-transitivity is by no means necessary for M to satisfy Separation, as we shall see later.) In this way we constructed, in Exercises 5.4 and 5.5, inner models HF, HC and HS consisting of hereditarily finite, hereditarily countable and hereditarily small sets, which respectively establish the independence of the axioms of Infinity, Power-set and Union. [Actually, to prove that Replacement holds in HS, we need to assume that the axiom of Choice holds in V, in order to show that the image of a small set under a function is small (cf. Lemma 8.1(ii)); but this is no hardship, since we shall shortly demonstrate the consistency of AC relative to ZF.]

Each of the inner models HF, HC and HS can be presented as the union of a 'hierarchy' like the von Neumann hierarchy, in which we replace the successor clause $V_{\alpha+1} = \mathscr{P}V_\alpha$ in the definition of the latter by, for example,

$$\mathrm{HF}_{\alpha+1} = \{x \in \mathscr{P}\mathrm{HF}_\alpha \mid x \text{ is finite}\}.$$

Thus we can regard them as 'slimmed-down' versions of the universe, in which the full power-set operation $\mathscr{P}$ has been replaced by something more restrictive. Another obvious way of 'cutting down' the universe is to truncate the von Neumann hierarchy at some ordinal α: if we stop at a successor ordinal α, the inner model V_α will fail to satisfy (at least) the Pair-set and Power-set axioms, but if we stop at a limit ordinal we may get something more interesting. In fact V_ω is simply another name for HF, as we saw in Exercise 6.4; but $V_{\omega+\omega}$ provides us with an independence proof for the axiom-scheme of Replacement. (Once again, we leave as an exercise the fact that $V_{\omega+\omega}$ satisfies all the other axioms of ZF; the reason why it fails Replacement is that the set (in V)

$$\{\langle n, \omega + n \rangle \mid n \in \omega\}$$

defines a function-class in $V_{\omega+\omega}$, whose domain is a set in $V_{\omega+\omega}$ but whose range is not.)

We cannot hope to demonstrate the independence of the axiom of Foundation using a standard model; but in fact Exercise 5.6 provides an independence proof for this axiom, using an interpretation where the underlying meta-set is interpreted as itself, but the interpretation of $\in$ is 'twisted' to yield a set which is a member of itself. On the other hand, the relative consistency of Foundation can be shown using a standard model: if V is a model for $\mathrm{ZF} - \{\mathrm{Fdn}\}$, then the class of regular sets in V (which we sometimes describe informally as 'the well-founded part of the universe') is a model of ZF. (Again, we leave the proof of this as an exercise; most of it follows straightforwardly from the result of Exercise 6.2.)

Our next goal is to prove the relative consistency of the axiom of Choice, i.e. ($\mathrm{Con}_{\mathrm{ZF}} \Rightarrow \mathrm{Con}_{\mathrm{ZF} \cup \{\mathrm{AC}\}}$). To do this, we shall follow Gödel's 1938 proof, the idea of which is again to build a standard model by 'slimming down' the von Neumann hierarchy. Given a set x, let $\mathscr{L}(x)$ denote the language of set theory enriched by adding a family of constants $\{\bar{y} \mid y \in x\}$; of course, we make V into a structure for this language by interpreting each $\bar{y}$ as y. We now define the set $\mathrm{Def}(x)$ of *definable* subsets of x to be the collection of sets of the form

$$\{y \in x \mid p(y)\}$$

where $p(y)$ is a formula of $\mathscr{L}(x)$ with one free variable y, and all quantifiers relativized to x. (Intuitively, if x is the collection of all sets we have constructed up to some point in a recursive construction of the universe, then $\text{Def}(x)$ is the collection of all those subsets of x which we are *immediately* forced to add if we want the axiom-scheme of Separation to hold. The reason why we allow members of x to appear as constants in the formula p is that they are possible values for the parameters in an instance of the Separation-scheme.) We now define the Gödel hierarchy $\alpha \mapsto L_\alpha$ by setting

$$L_0 = \varnothing$$
$$L_\alpha = \text{Def}(L_\beta) \qquad \text{if } \alpha = \beta^+$$
$$L_\alpha = \bigcup \{L_\beta \mid \beta < \alpha\} \quad \text{if } \alpha \text{ is a limit ordinal.}$$

We define the class L of *constructible* sets to be the union of all the L_α, $\alpha \in \text{ON}$.

The Gödel hierarchy grows much more slowly than the von Neumann hierarchy; the two agree for $\alpha \leqslant \omega$, since any finite subset of a set x is in $\text{Def}(x)$, but $L_{\omega+1}$ is countable (since $\mathscr{L}(L_\omega)$ is) and hence cannot be the whole of $V_{\omega+1}$. (In fact L_α is countable for every countable ordinal α.) On the other hand, the L_α fit together rather well. We note first that if x is transitive then $x \subseteq \text{Def}(x)$ (since if $z \in x$ we have $z = \{y \in x \mid y \in \bar{z}\} \in \text{Def}(x)$); and an easy induction shows that each L_α is transitive. Also, if x has a well-ordering then so does the set of primitive symbols of $\mathscr{L}(x)$; hence we can well-order the set of formulae of $\mathscr{L}(x)$ with one free variable (e.g. lexicographically), and so we obtain a well-ordering of $\text{Def}(x)$. If x is both well-ordered and transitive, then it is easy to arrange things so that x occurs as an initial segment of $\text{Def}(x)$ (with its original well-ordering); hence we can recursively define a well-ordering of each L_α, and indeed of the entire constructible universe L.

Proposition 9.3. If V is a model of ZF, then L is a model of
ZF $\cup \{\text{AC}\}$.

Proof. The remarks in the preceding paragraph ensure that, provided L is a model of ZF, it will also satisfy AC. The verification in L of most of the axioms of ZF is straightforward; the two non-trivial ones are Separation and Power-set. We discuss the latter first. Note that L is not necessarily super-transitive; so we are not trying to prove that the full power-set $\mathscr{P}_V x$ of a constructible set x is constructible, but rather that

$$\mathscr{P}_L x = \{ y \in \mathscr{P}_V x \mid y \in L \}$$

is a member of L. (On the other hand, $\mathscr{P}_L x$ will in general be larger than $\mathrm{Def}(x)$: for example, $L_{\omega + 1} = \mathrm{Def}(L_\omega)$ is countable even in L (i.e. there is a bijection $\omega \to L_{\omega + 1}$ which is a set in L), and so cannot be the power-set of L_ω in L.) However, for each $y \in \mathscr{P}_L x$ there is an ordinal α such that $y \in L_\alpha$, and so by Replacement (in V) we can find an ordinal γ which is an upper bound for all such α. Now, if we rewrite the definition of $\mathscr{P}_L x$ as

$$\{ y \in L_\gamma \mid (\forall z \in L_\gamma)((z \in y) \Rightarrow (z \in \bar{x})) \},$$

we see that it belongs to $L_{\gamma + 1}$.

The verification of the Separation-scheme is similar. Let x be a constructible set, and p a formula with one free variable y (the other free variables having been replaced by constants corresponding to elements of L). We must show that $x \cap [p]_L$ is a set in L. Now if α is an ordinal large enough for L_α to contain x and all the sets appearing as constants in p, then we have

$$x \cap [p]_{L_\alpha} \in \mathrm{Def}(L_\alpha) = L_{\alpha + 1};$$

the problem is that, if p contains quantifiers, it is not necessarily true that $[p]_L \cap L_\alpha = [p]_{L_\alpha}$ (cf. Exercise 3.3). We must show that, given p, there exists β such that $[p]_L \cap L_\beta = [p]_{L_\beta}$; we do this by induction on the structure of p, using the validity of Replacement in V in the case when p has the form $(\forall z)q$. $\square$

The model L also satisfies Gödel's *Axiom of Constructibility*, which is the assertion that every set is constructible (in symbols, $V = L$); i.e., if we perform the construction of the L_α within L, we end up with the whole of L. (In fact a stronger result is true: for a given V, L is the unique smallest substructure of V which contains all the ordinals of V and satisfies the axioms of ZF.) By what we have just shown, the axiom $V = L$ implies AC; it also settles the status of a number of other assertions which cannot be either proved or disproved in ZF, including the Generalized Continuum Hypothesis. [GCH is true in L; intuitively, the reason for this is that a counterexample to GCH – for example, a subset of $\mathscr{P}\omega$ whose cardinality is strictly between that of ω and that of $\mathscr{P}\omega$ – could not possibly be constructible, but the actual proof of GCH in L is rather more complicated.]

It remains to discuss the independence of AC and of GCH. The first 'independence proof' for AC was given by A. Fraenkel in 1922; however, the model he constructed was not a model of ZF, in that it contained 'atoms' ('Urelemente' in German) which were not sets (and in particular had no members) but were allowed to be members of sets. (It is intuitively clear that if we adjoin enough atoms to our universe, and make them sufficiently indistinguishable, then we are going to have difficulty in finding choice functions for sets of sets of atoms. The indistinguishability is ensured by allowing some group of permutations G to act on the atoms, and then taking as elements of the model only those sets which are 'nearly' G-invariant.) Fraenkel's model thus fails to satisfy the axiom of Extensionality, as we have formulated it.

In 1939, A. Mostowski modified Fraenkel's method so that Extensionality was preserved, although Foundation was violated (he replaced the atoms by sets x satisfying $x = \{x\}$, which also tend to be indistinguishable). Also, by taking a linearly ordered set of such xs and allowing only order-preserving permutations of them, he showed that 'Every set can be linearly ordered' can remain true even when AC fails; later, J. D. Halpern showed that Mostowski's model in fact satisfies the Prime Ideal Theorem (cf. Exercise 7.6). It was not until 1963 that P. Cohen developed a technique for 'forcing' the counterexample to AC which exists in Mostowski's model into the well-founded part of the universe, and hence (by taking the well-founded part of the resulting model) obtaining a model of ZF in which AC fails. (Actually, Cohen's independence proof for AC was originally presented in a rather different way, but it's equivalent to the technique just described.) Cohen's 'forcing' technique (which is a technique for adjoining things to a model of set theory – in the case just cited, what we need to adjoin is an injection from a given non-well-orderable set to (say) $\mathscr{P}\omega$) also enabled him to produce a model of ZF (and of AC) in which the Continuum Hypothesis fails, as we mentioned at the end of Chapter 8. Since 1963 the technique has been greatly developed to prove other independence results, which we do not have space to discuss here.

Index of definitions

Index of names